EYEWITNESS 👁 SCIENCE

ELECTRICITY

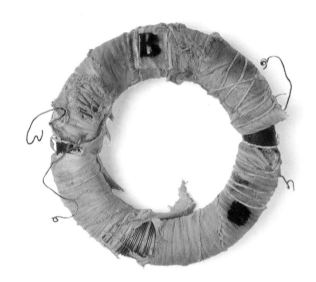

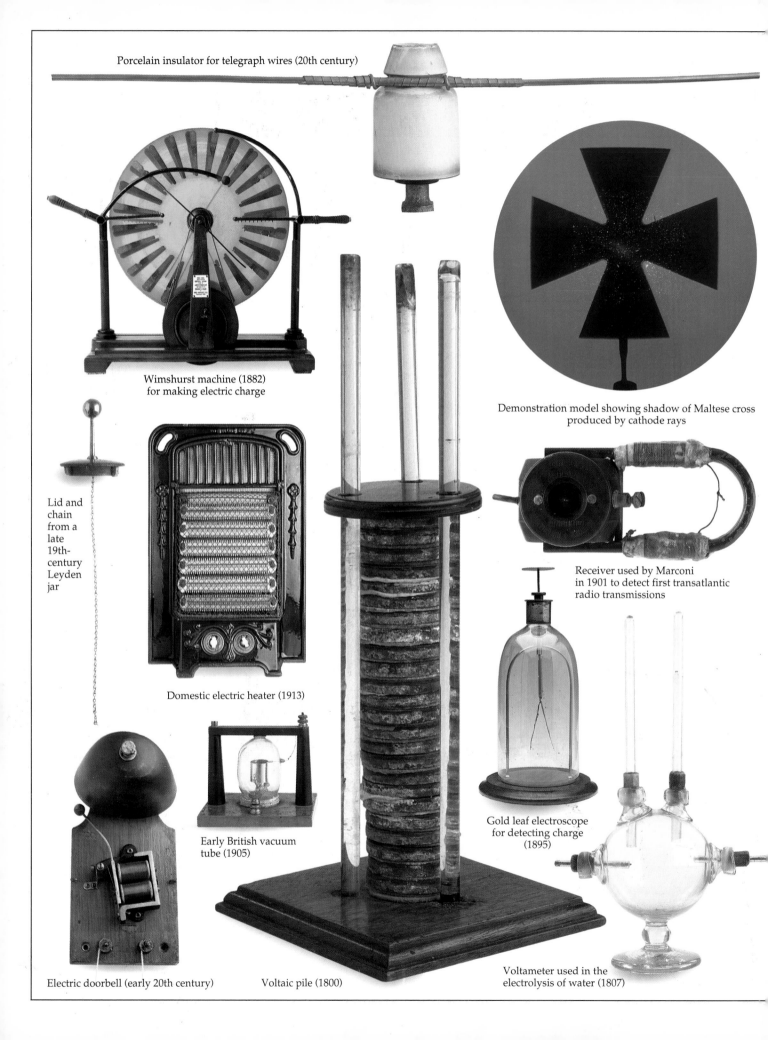

Porcelain insulator for telegraph wires (20th century)

Wimshurst machine (1882) for making electric charge

Demonstration model showing shadow of Maltese cross produced by cathode rays

Lid and chain from a late 19th-century Leyden jar

Domestic electric heater (1913)

Receiver used by Marconi in 1901 to detect first transatlantic radio transmissions

Early British vacuum tube (1905)

Gold leaf electroscope for detecting charge (1895)

Electric doorbell (early 20th century)

Voltaic pile (1800)

Voltameter used in the electrolysis of water (1807)

Ammeter

Voltmeter

EYEWITNESS  SCIENCE

ELECTRICITY

Written by
STEVE PARKER

Pure potassium produced by electrolysis
by Humphry Davy (1807)

Early telegraph cables encased in wood (1837)

Making a wire
glow red hot using
a battery (1895)

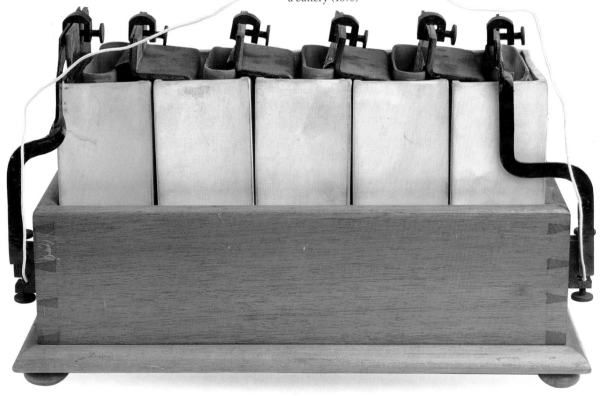

Stoddart

Telephone handset (1895)

A DORLING KINDERSLEY BOOK

☞ NOTE TO PARENTS AND TEACHERS
The **Eyewitness Science** series encourages children to observe and question the world around them. It will help families to answer their questions about why and how things work – from daily occurrences in the home to the mysteries of space. In school, these books are a valuable resource. Teachers will find them useful for work in many subjects, and the experiments and demonstrations in the books can ☞ serve as an inspiration for classroom activities.

Project Editor Charyn Jones
Senior Art Editor Neville Graham
Design Assistant Marianna Papachrysanthou
DTP Manager Joanna Figg-Latham
Production Eunice Paterson
Managing Editor Josephine Buchanan
Special Photography Clive Streeter
Editorial Consultant Neil Brown, Science Museum, London
US Editor Charles A. Wills
US Consultant Harvey B. Loomis

This Eyewitness ®/™Science book first published in Great Britain in 1992 by Dorling Kindersley Limited, 9 Henrietta Street, London WC2E 8PS

First published in Canada in 1992 by Stoddart Publishing Co. Limited 34 Lesmill Road, Toronto, Canada M3B 2T6

Canadian Cataloguing in Publication Data
Parker, Steve
Electricity
(Eyewitness science)
ISBN 0-7737-2613-6

1. Electricity - Juvenile Literature. I. Title.
II. Series.
QC527.2.P37 1992   J537   C92-093785-3

Reproduced by Colourscan, Singapore
Printed and bound in Italy by A. Mondadori Editore, Verona

Damaged lightning rod (1916)

10,000 volt electricity cable insulated with waxed paper (1890)

Interior carbon arc lamp (1870-1880)

Zinc, pasteboard, and copper discs from a voltaic pile (1800)

Pith ball electrometer (19th century)

G. ADAMS
LONDON

Contents

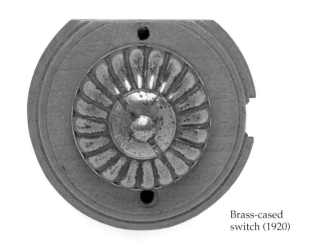

Brass-cased
switch (1920)

A mysterious force

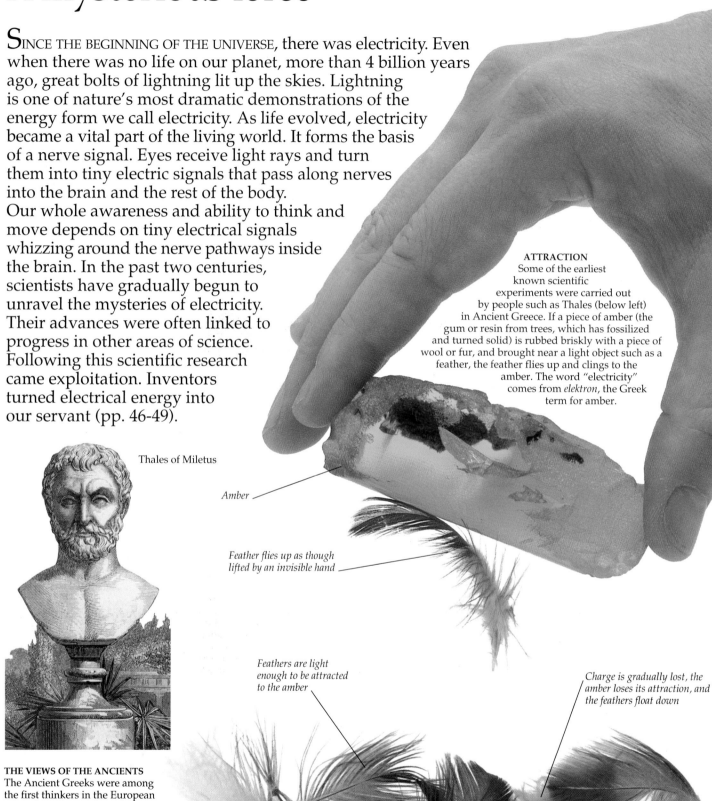

Since the beginning of the universe, there was electricity. Even when there was no life on our planet, more than 4 billion years ago, great bolts of lightning lit up the skies. Lightning is one of nature's most dramatic demonstrations of the energy form we call electricity. As life evolved, electricity became a vital part of the living world. It forms the basis of a nerve signal. Eyes receive light rays and turn them into tiny electric signals that pass along nerves into the brain and the rest of the body. Our whole awareness and ability to think and move depends on tiny electrical signals whizzing around the nerve pathways inside the brain. In the past two centuries, scientists have gradually begun to unravel the mysteries of electricity. Their advances were often linked to progress in other areas of science. Following this scientific research came exploitation. Inventors turned electrical energy into our servant (pp. 46-49).

ATTRACTION
Some of the earliest known scientific experiments were carried out by people such as Thales (below left) in Ancient Greece. If a piece of amber (the gum or resin from trees, which has fossilized and turned solid) is rubbed briskly with a piece of wool or fur, and brought near a light object such as a feather, the feather flies up and clings to the amber. The word "electricity" comes from *elektron*, the Greek term for amber.

Thales of Miletus

Amber

Feather flies up as though lifted by an invisible hand

Feathers are light enough to be attracted to the amber

Charge is gradually lost, the amber loses its attraction, and the feathers float down

THE VIEWS OF THE ANCIENTS
The Ancient Greeks were among the first thinkers in the European scientific tradition. One of the earliest Greek scientists was Thales of Miletus (*c.* 625-547 BC), a skillful mathematician. None of Thales' own writings survives, but reports of his work show that he probably carried out simple experiments into the effects of what we now call electricity and magnetism.

ELECTRICITY IN THE AIR

The lightning bolt is an awesome example of electricity in action. The explanations of what electricity is, where it comes from, and how it works, are central to our understanding of matter and the fundamental forces of nature. Lightning is the result of the discharge of an electric charge in a cloud. The energy of the discharge is so great that it produces an intense trail of light, heat, and sound – thunder. It can destroy buildings, kill humans, and consume trees in a sheet of flame. Observations of lightning led scientists and thinkers such as Benjamin Franklin (pp. 8-9) to investigate and begin to unravel the mysteries of electric charge.

PLAYING TRICKS

In times gone by, teachers showed the attraction of magnetic and electrically charged objects, even though they did not understand their nature. Magicians still use the attracting powers of magnetism and "static" electricity (or electric charge) in their acts.

ELECTRICITY IN THE BODY

Human life depends on electricity. About every second, tiny electric signals spread through the heart muscle, triggering and coordinating a heartbeat. These signals send "echoes" through the body tissues to the skin. Here they can be detected by metal sensors and displayed as a wavy line called the electrocardiogram (p. 51).

ELECTRICITY IN ANIMALS

In the animal world, a busy muscle produces small pulses of electricity. Creatures have capitalized on this for hunting and killing. The electric ray has modified muscle blocks on either side of its head. These jellylike "living batteries" send shock waves through the water to stun or kill a nearby victim. A typical shark has 1,000 electricity-sensing pits on its skin, mainly around its head. They pick up tiny electric pulses from the active muscles of fish. In complete darkness a shark can home in on its meal with unerring accuracy, using its electrical-navigation sensors.

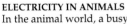

Electricity-producing muscle blocks

Nerves controlling electric organs

Atlantic torpedo or electric ray

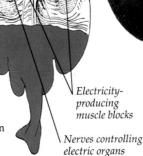

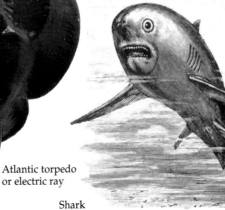

Shark

Ideas about electricity

I**N THE EARLY DAYS** of experiments with electricity, scientists had no batteries to make electricity. Instead, they made it themselves by rubbing certain materials together. Around 1600 William Gilbert suggested there were two kinds of electricity, based on the material used to do the rubbing, though he did not know why this was so. Glass rubbed with silk made vitreous electricity, and amber rubbed with fur made resinous electricity. His experiments showed that objects containing the same kind of electricity repelled each other, while those containing different kinds attracted each other (pp. 10-11). Benjamin Franklin also believed that there were two kinds of electricity (pp. 20-21). He proposed that electric charge was like "fluid" spreading itself through an object. It could jump to another object, making a spark.

Versorium

Thin, finely balanced pointer of light wood

Upright rod

GILBERT'S VERSORIUM
William Gilbert (1544-1603) was a doctor to Queen Elizabeth I of England. In 1600 he wrote about the mysterious forces of magnetism and electricity in his book *De Magnete* (*On the Magnet*). Gilbert was the first person to use the word "electric," and he invented possibly the earliest electrical instrument, the versorium. Objects like paper and straw, which had an electric charge when rubbed, made the pointer of the versorium swing toward them. He called these attracting substances "electrics." Those that did not attract the pointer were "nonelectrics."

Tip bent by intense heat

Lightning rod

DIRECT HIT
A lightning bolt is a giant spark of electric charge jumping from a thunder-cloud to the ground. If lightning hits anything, it burns it. A rod directs charge through a copper strip to the earth, leaving the building safe.

A RISKY EXPERIMENT
The lightning experiments of Benjamin Franklin (1706-1790) were copied by other investigators. The scientist Georg Richmann at St. Petersburg, Russia, was killed by the electric charge of lightning when he held a wire-tipped pole up high in a storm. The sparks from lightning are in fact similar to those obtained in laboratories.

BENJAMIN FRANKLIN
To "collect" electricity from lightning, Franklin flew a kite in a thunderstorm. Sparks jumped when the lightning flowed along a kite string to which a key had been tied.

Charge-collecting metal combs

Charge conducted down chains to jars

Leyden jar

Outer metal coating of Leyden jar

Brush touches across disc as part of charge buildup

Discharging spheres

Wimshurst machine

As the charge builds up, a spark leaps across the small gap

Conductor arms

PRODUCING ELECTRIC CHARGE

For many years, the Wimshurst machine was used to produce electric charge. The machine works by induction (p. 11). Turning the crank handle starts a process involving the metal sectors which are stuck on the outside of two glass discs – one wheel rotates in the opposite direction to the other – and metal combs that point to the discs, but do not actually touch them. This process multiplies small stray electric charges many times. The charge produced is stored in Leyden jars. When enough charge has built up, it jumps between the two discharging spheres, producing a bright spark. The device was developed by James Wimshurst (1832-1903). Wimshurst machines were used to demonstrate making static electricity – as the charge is sometimes called – until at least the 1960s. The largest machine had discs that were 7 feet (2.12 m) in diameter.

James Wimshurst

Leyden jar being discharged

Metal sectors stuck to outside of both discs

Two contra-rotating glass discs placed close together

Metal combs point to sectors on glass discs and collect charge

Metal ball

Discharger rod

Spark

Lid

Glass jar

Outer metal coating

Pulley

Crossed drive belt

Drive belt

Metal ball

Lid

Leyden jar with lid and chain

STORING CHARGE

The Leyden jar was an early device scientists used for storing the electric charge they were making. It is named after the place where it was developed in 1746, the University of Leyden in the Netherlands. The electric charge flows down the metal chain to metal coating inside the jar. It cannot leak away through the glass jar, so it builds up. If the discharger rod is held near the jar (above), the charge leaps from the ball on top, through the discharger to the outer metal coating on the jar – causing a spark. The Leyden jar was an early type of capacitor or condenser (pp. 12-13).

Metal chain

Glass jar

Leyden jar

Hand-turned crank handle

Drive pulley

Outer metal coating

The quest for knowledge

DURING THE 18TH CENTURY many scientists experimented with electric charge in their laboratories. As yet, there were no practical uses for electricity; what interested scientists was the quest for knowledge. They observed how electric charge could be seen as sparks, and how it behaved differently with different substances. Since electricity was invisible, instruments were needed to detect and measure it. Initially, progress was haphazard. There was no way to make a sustained flow of electric charge – that came later from the battery (pp. 16-19). Startling new discoveries were made that are now taken for granted. For instance, in the 1720s the English scientist Stephen Gray (1666-1736) proposed that any object which touches an "electrified (charged) body will itself become electrified." Charge transferring from one substance to another one that touches it is a process called electrical conduction.

JOSEPH PRIESTLEY
Before achieving fame as a chemist and the discoverer of oxygen, Priestley (1733-1804) was interested in electricity. In 1767 he published the earliest history of electrical science – *The History and Present State of Electricity*. This was his personal assessment of contemporary studies.

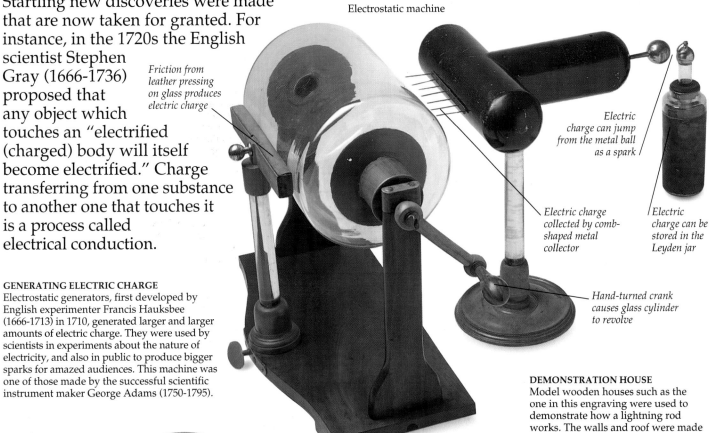

Electrostatic machine

Friction from leather pressing on glass produces electric charge

Electric charge can jump from the metal ball as a spark

Electric charge collected by comb-shaped metal collector

Electric charge can be stored in the Leyden jar

Hand-turned crank causes glass cylinder to revolve

GENERATING ELECTRIC CHARGE
Electrostatic generators, first developed by English experimenter Francis Hauksbee (1666-1713) in 1710, generated larger and larger amounts of electric charge. They were used by scientists in experiments about the nature of electricity, and also in public to produce bigger sparks for amazed audiences. This machine was one of those made by the successful scientific instrument maker George Adams (1750-1795).

DEMONSTRATION HOUSE
Model wooden houses such as the one in this engraving were used to demonstrate how a lightning rod works. The walls and roof were made to fall apart easily. Hidden inside was a container of gunpowder. Electric charge from an electrostatic generator, representing lightning, would normally be guided down into the ground by a metal lightning rod. If this was disconnected in the model, the charge created a spark which set off the gunpowder, blowing off the roof so the walls fell flat.

Horse and rider electrostatic toy

Charge sprays off the end of the point

Vanes spin

Connection to source of charge

Rider and horse go round

MOVING TOYS
Electric charge was used to work this country scene from the mid-19th century. The metal point was connected to a source of charge; the charge "sprayed" off the end of the point and jumped to the vanes, moving them and the model around.

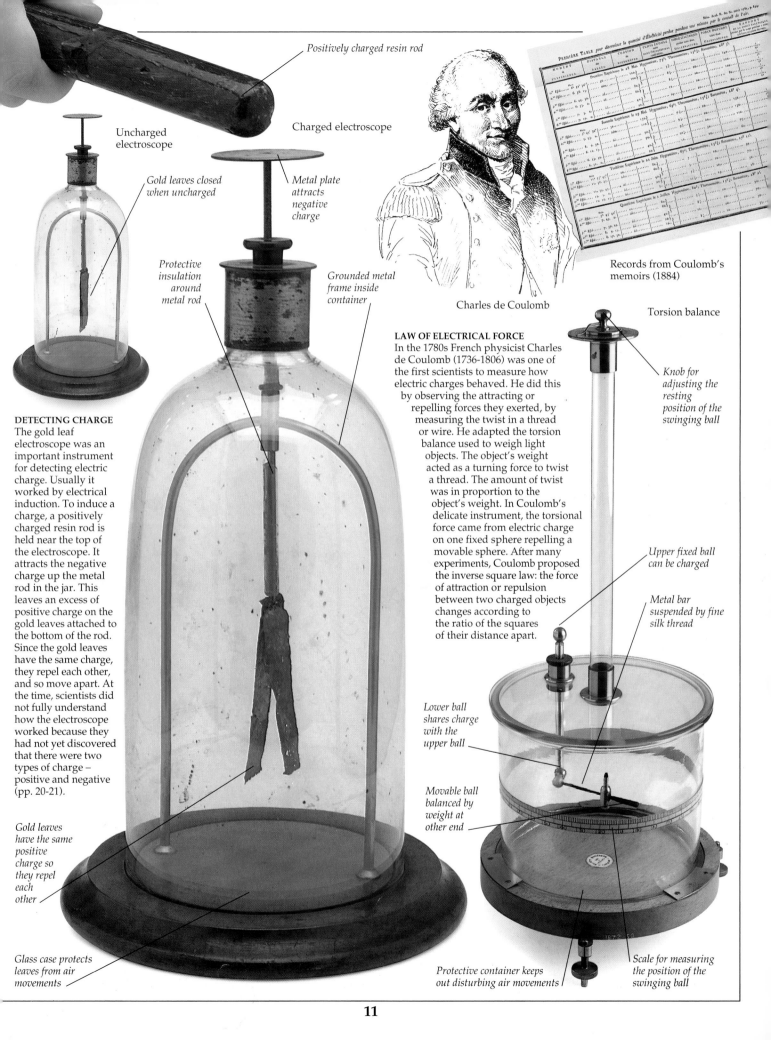

Positively charged resin rod

Uncharged electroscope

Gold leaves closed when uncharged

Charged electroscope

Metal plate attracts negative charge

Protective insulation around metal rod

Grounded metal frame inside container

Records from Coulomb's memoirs (1884)

Charles de Coulomb

Torsion balance

Knob for adjusting the resting position of the swinging ball

DETECTING CHARGE
The gold leaf electroscope was an important instrument for detecting electric charge. Usually it worked by electrical induction. To induce a charge, a positively charged resin rod is held near the top of the electroscope. It attracts the negative charge up the metal rod in the jar. This leaves an excess of positive charge on the gold leaves attached to the bottom of the rod. Since the gold leaves have the same charge, they repel each other, and so move apart. At the time, scientists did not fully understand how the electroscope worked because they had not yet discovered that there were two types of charge – positive and negative (pp. 20-21).

LAW OF ELECTRICAL FORCE
In the 1780s French physicist Charles de Coulomb (1736-1806) was one of the first scientists to measure how electric charges behaved. He did this by observing the attracting or repelling forces they exerted, by measuring the twist in a thread or wire. He adapted the torsion balance used to weigh light objects. The object's weight acted as a turning force to twist a thread. The amount of twist was in proportion to the object's weight. In Coulomb's delicate instrument, the torsional force came from electric charge on one fixed sphere repelling a movable sphere. After many experiments, Coulomb proposed the inverse square law: the force of attraction or repulsion between two charged objects changes according to the ratio of the squares of their distance apart.

Upper fixed ball can be charged

Metal bar suspended by fine silk thread

Lower ball shares charge with the upper ball

Movable ball balanced by weight at other end

Gold leaves have the same positive charge so they repel each other

Glass case protects leaves from air movements

Protective container keeps out disturbing air movements

Scale for measuring the position of the swinging ball

11

Collecting electric charge

Today, charge-producing electrostatic generators are an unfamiliar sight, confined to museums and research laboratories. These machines were designed to produce large charges and extremely high voltages. Charge-storing devices are vital components found inside many electrical devices, from washing machines to personal stereos. There are many different types of condenser (opposite), yet they all use the same principle as the Leyden jar (pp. 8-9), which was the first electrical storage container. Condensers, sometimes called capacitors, are the only electric devices, other than batteries, that can store electrical energy. They are also used to separate alternating current (AC) from direct current (DC).

VAN DE GRAAFF GENERATOR

American scientist Robert Van de Graaff (1901-1967) developed a machine in the early 1930s for collecting and storing huge amounts of electric charge. He is shown here with one of his smaller electrostatic generators. The charge in the machine, which is named after him, builds up on a metal sphere, and reaches an incredible 10 million volts. The machine is mostly used as a research tool for studying the particles that make up atoms. The vast energy represented by its accumulated charge is transferred to the atoms' particles by accelerating the particles to enormous speed, so that scientists can study their interactions as they smash together.

Robert Van de Graaff

ACCUMULATING CHARGE

In the Van de Graaff generator (below), the source of positive electric charge is a comb-shaped charge-sprayer connected to the electricity supply. The charge is carried on a moving belt to a charge collector above. This transfers it to the exterior of the large metal sphere. The sphere is mounted on a column that prevents the charge from leaking away. In atomic research there are no sparks; the fast-moving sub-atomic particles carry the charge away. In public demonstrations such as this model at the Boston Museum in Massachusetts (right), the accumulated charge leaps across to another piece of metal nearby, with a giant spark like a mini-lightning bolt.

THE SPARK OF LIFE

The electrostatic generator's giant sparks made impressive special effects at the movies. Their similarity to lightning bolts fitted neatly into the plot of Mary Shelley's horror story *Frankenstein*. Doctor Frankenstein's monster, made from sewn-together bits of dead bodies, is jolted into life by the shock from a lightning bolt.

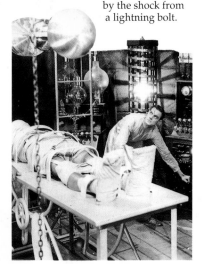

SPEEDING PARTICLES

The Cockcroft-Walton generator at Brookhaven Laboratory on Long Island, New York, generates electrical energy in the form of ultra-high voltages. The energy speeds up bits of atoms so fast that they travel a distance equivalent to the moon and back in four seconds.

How the Van de Graaff works

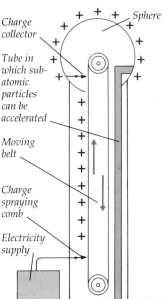

Charge collector

Sphere

Tube in which sub-atomic particles can be accelerated

Moving belt

Charge spraying comb

Electricity supply

Condensers

There are many different types of condenser that store electric charge. Two sheets of metal, or a similar electrical conductor, are separated by an insulating material such as paper or air. Capacitance is measured in units called farads, after Michael Faraday (pp. 34-35). One farad is a huge amount of charge, and most modern condensers are fingertip-sized items rated at microfarads (millionths of a farad) or less.

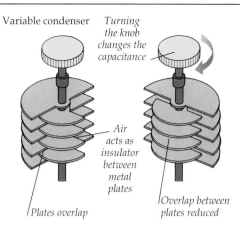

Variable condenser

Turning the knob changes the capacitance

Air acts as insulator between metal plates

Plates overlap

Overlap between plates reduced

CONDENSERS AT WORK
In electric devices, a condenser is used to store charge and release it at a certain moment, or to resist the continuous flow of charge through it. Bigger conducting plates, and a narrower gap between them, increase the amount of charge stored. The tuning knob on this 1950s radio operates a variable condenser, altering the overlap between its stack of metal plates. This changes the capacitance and causes the radio to respond to signals of a different frequency, that is, to pick up a different radio station.

Metal plates of variable condenser

Tuning knob

Using charges

THE DUST RESTING on a television screen is an example of electrostatic attraction. The glass surface of the screen becomes electrically charged while the television is on. It then attracts and holds any floating specks which happen to come near. This phenomenon of electrostatic forces, where there is attraction by unlike charges, and repulsion by like ones, is put to work in a variety of modern machines and processes. For instance, in the body-painting shop of a car manufacturer, tiny droplets of spray paint are all given the same electric charge. They repel each other and are attracted towards the car's body, and so settle on it as a more even coating. This is exactly the same principle as the charge that amber, when rubbed, produces to pick up feathers (pp. 6-7).

THE CARLSON COPIER
In 1938 American lawyer Chester Carlson (1906-1968) devised a process known as electrophotography. He wanted a machine that could duplicate patent application forms – not only the words, but also complicated drawings. He invented an electrostatic printer, or xerography machine, (from the Greek words *xeros* meaning "dry," and *graphos* for "writing.") The first xerographic print (above) was made by Carlson in 1938, but the first commercial copies were not produced until the 1950s.

Carlson's first print

Carlson with his early copier

Early copier (1960)

Charging chamber with top removed

Outer casing

Selenium-coated plate

Charging wires

Plate-charging button

Edge of selenium-coated plate

Developing tray with toner inside

Special toner tray (used for copying half-tone pictures)

"DRY WRITING"
The early copier uses the attraction of unlike electric charges (pp. 10-11). At its heart is a special metal plate coated with a substance called selenium. A pattern of positive charges on the plate, representing the black areas to be copied, attracts negatively charged particles of a fine black powder – the toner. (The toner becomes negatively charged by contact with tiny glass beads in the developing tray.) The toner pattern is transferred to a blank sheet of paper and heat-sealed in place. In a modern copier, the selenium-coated plate is on a rotating drum. Otherwise the process is much the same as this 1960s machine, but now it happens automatically.

TONE TRAY

Making a
photocopy

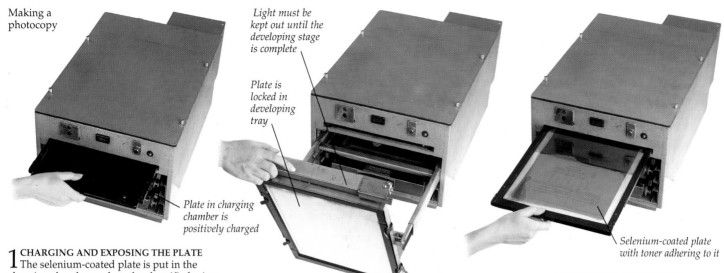

Light must be kept out until the developing stage is complete

Plate is locked in developing tray

Plate in charging chamber is positively charged

Selenium-coated plate with toner adhering to it

1 CHARGING AND EXPOSING THE PLATE

The selenium-coated plate is put in the charging chamber and, as the electrified wires pass over, it receives an even coating of positive charges. The protective shield is replaced and the plate removed to a camera, where it is exposed to the document to be copied. An image of the document is shone on to it by a camera lens. Where light hits the plate, from the white areas of the original document, the selenium becomes a conductor, and the charge flows away. Where no light reaches, the charge remains.

2 DEVELOPING THE PLATE

The plate now has an exact mirror copy of the original document on its surface, in the form of a pattern of positive electrostatic charge. With the protective shield in place the plate is locked to the developing tray. The shield is removed, and the plate swung backward and forward. As this happens the toner in the developing tray cascades over the plate. Its tiny negatively charged grains are attracted to the positively charged areas of the plate, where they stick.

3 THE DEVELOPED IMAGE

The plate is removed from the developer tray. This reveals the selenium-coated plate with the fine powder adhering to it, as an exact but mirror-image replica of the original document. The plate is returned to the charging chamber.

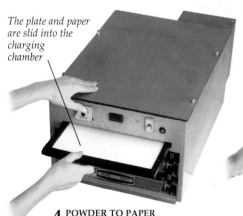

The plate and paper are slid into the charging chamber

The paper now holds a duplicate of the original

The paper is baked in the fuser

4 POWDER TO PAPER

The next stage also depends on electrostatic attraction. A sheet of blank paper is placed over the plate and its powder image. The plate and paper are pushed back into the charging chamber and withdrawn again while the transfer switch is depressed. The paper becomes positively charged so that it attracts the toner powder away from the plate.

5 REMOVING THE COPY

The paper is carefully lifted from the plate, bringing with it the pattern of toner, which is now stuck to it by electrostatic attraction. The copied parts have been reversed again, so that they are an exact duplicate of the original.

6 HEAT-SEALING THE COPY

Finally, the plain paper with its powder pattern is placed on the fuser tray and pushed into an oven-like chamber for a few seconds. The powder bakes and melts into the fibers of the paper, permanently sealing the image. The entire process takes about three minutes – much longer than the couple of seconds in a modern photocopier.

DUST-EATING AIR FILTER

The electrostatic air filter on this 1930s cigarette card, showing a man demonstrating the filter with cigarette smoke, uses a fan to draw in a stream of dusty, impure air. A prefilter traps the bigger floating particles. The remaining small ones pass through the first electrified grid of a device called the electrostatic precipitator. This grid gives each particle a negative charge. The particles are repelled from the negative wires onto the precipitator's second, positively charged grid. They are attracted to it, and stick to its mesh. A filter then absorbs any odors and cleaned air blows out the other end.

How an electrostatic precipitator works

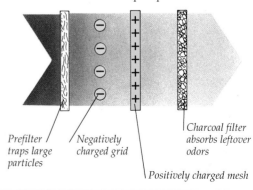

Prefilter traps large particles

Negatively charged grid

Charcoal filter absorbs leftover odors

Positively charged mesh

A flow of charge

In 1780, while Italian anatomist Luigi Galvani (1737-1798) was dissecting and studying a frog, he noticed that when his sharp scalpel touched the nerves in the frog's leg, the leg twitched. Galvani suspected that there was electricity in the frog's muscles, and experimented to explain what he had seen. Soon another Italian, Alessandro Volta (1745-1827), heard of the incident. Volta disagreed with Galvani's ideas. He had already developed a device for producing small amounts of electric charge – the electrophorus. In 1800 he announced that he had found a new source of electricity, one which produced electricity continuously (unlike the Leyden jar, which discharged instantly). It became known as the voltaic pile. Galvani thought electricity came from animals' bodies when touched by two metals, calling it "animal electricity." Volta thought it came from contact between metals only; he called it "metallic electricity." Galvani and Volta disagreed strongly, and they and their supporters argued for years. We now know that neither was totally correct.

Voltaic pile

Glass supporting rod

THE VOLTAIC PILE
This device used two different metals, separated by moist chemicals, to produce a flow of electric charge. The original voltaic pile used three types of disc: zinc, pasteboard or leather, and copper. The pasteboard was soaked in a solution of salt or weak acid such as vinegar. In an electrochemical reaction, the copper loses electrons (p. 20) to the solution and the zinc gains electrons from the solution. At the same time, zinc dissolves and hydrogen gas is produced at the surface of the copper. When the charge flows away along wires, the chemicals separate out more charge. And so electric charge continues to flow. Volta had invented the earliest electric cell. He piled up many of these cells or three-disc units to strengthen the effect. In this way he produced the first battery, which is a collection of cells (pp. 18-19).

Copper disc

Zinc disc

Pasteboard disc

Wooden base

Galvani's laboratory

LUIGI GALVANI
Galvani studied the effects of electric charge from Leyden jars and electrostatic machines on animals. In a number of ways Galvani brought animal tissue into contact with two metals in his laboratory (above). He noticed convulsions in the limbs of the dead animals. Chemicals in the nerves and muscles, when placed between the metals, had caused an electrochemical reaction (pp. 18-19), making an electric cell.

Luigi Galvani

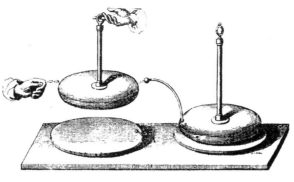

Volta's letter
to the Royal
Society (1800)

Alessandro
Volta

THE ELECTROPHORUS

Volta invented the electrophorus in 1775. It consisted of a metal plate, which could be charged, placed on an insulated base. With a handle, which was also made from an insulating material, the electrophorous could be used as a portable carrier of electric charge, to transfer charge to another apparatus, or further charge a Leyden jar. The process of charging and discharging the electrophorus could be repeated many times.

STORING CHARGE

The insulated base of the electrophorus, which was made from a resinous substance, was negatively charged by rubbing it so that when the metal plate was placed on top of it, the bottom of the disc acquired a positive charge, while the top was negatively charged. The negative charge was drained away by touching the plate (above) and the positively charged plate could then be taken away, and the charge on it used.

Glass handle

Brass disc carries the charge

Ebonite base which can be charged

ALESSANDRO VOLTA

On 20 March 1800 Volta sketched a U-shaped voltaic pile in a letter to Sir Joseph Banks, president of the Royal Society of London. The pile was the first practical battery. The modern unit of electric potential, which is the strength or "electric push" of the flowing charge, is the volt (pp. 24-25), named in honor of Volta. His apparatus produced just over one volt for every set of discs. The electric ray (p. 7) can send a 100-volt shock through the water. A modern house circuit carries 110 or 220 volts and high-voltage power lines carry up to half a million volts or more.

Sets of discs

Insulated wire

Wire inserted under disc

WIRING UP THE BATTERY

To connect up the voltaic pile to a piece of equipment, two insulated copper wires were attached to sets of discs. With 24 sets of discs this battery produced about 24 volts, which Volta detected with the tip of his tongue on the connecting wires.

Components of a voltaic cell

Zinc disc

Salt-soaked pasteboard disc

Copper disc

Copper disc

Zinc disc

Electricity from chemicals

RESEARCHERS FOUND THAT THE SIMPLEST electricity-making unit, or electric cell, was two plates of different metals in a jar filled with liquid. The metal plates are called electrodes. They are conductors through which electricity can enter or leave. The positive electrode is called the anode and the negative electrode is called the cathode. The liquid, which must be able to conduct electricity, is called the electrolyte. Several cells joined together form a battery. There have been many types and sizes of cells and batteries. Some of them used strong acids or other noxious chemicals as the electrolyte. The first batteries supplied electricity for research, in the laboratory. Large numbers of batteries were used for electric telegraphs (pp. 56-57). An important advance was the "dry" cell, a development of the Leclanché cell (below), which uses a jellylike paste instead of liquid. More recent advances include alkaline and other long-life cells.

SECONDARY CELLS
Secondary or rechargeable batteries were developed in 1859 by French scientist Gaston Planté (1834-1889). In a primary cell, like a flashlight battery, the chemical reactions eventually become spent and no longer produce electricity. In a secondary cell, the reactions can be reversed by electricity from another source. After recharging, the cell produces current again.

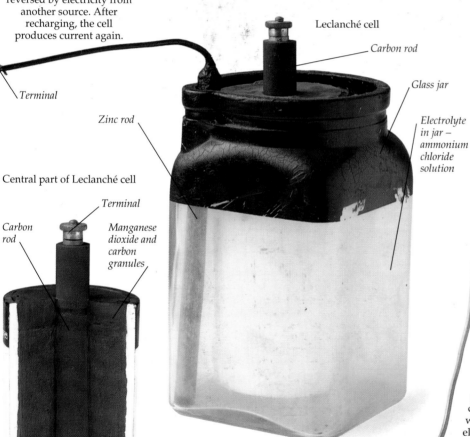

Leclanché cell

Carbon rod

Glass jar

Electrolyte in jar – ammonium chloride solution

Terminal

Zinc rod

Central part of Leclanché cell

Terminal

Carbon rod

Manganese dioxide and carbon granules

LECLANCHE CELL
During the 1860s French chemist Georges Leclanché (1838-1882) devised a cell in which one electrode was a zinc rod, and the other was a carbon rod inside a pot of manganese dioxide and carbon granules. Between was a solution of ammonium chloride. The cell produced about 1.5 volts. It did not contain dangerous acid, and it soon became a popular and relatively portable electricity-maker, and the forerunner of the flashlight battery.

Zinc cathode

Corrosion-resistant glass container

THE CELL AT WORK
A basic electrical cell has copper and zinc plates immersed in sulfuric acid. When the plates are connected by a conducting wire, chemical reactions occur. Hydrogen gas is given off at the copper plate, which loses electrons (p. 33) to the solution, becoming positively charged. Zinc dissolves from the zinc plate, leaving behind electrons which make the plate negative. The electrons move through the wire from the zinc plate towards the copper. This constitutes the electric current which continues until the zinc is eaten away or the acid used up.

MASSIVE BATTERY
Humphry Davy (1778-1829) was professor at London's Royal Institution. In 1807 he used a roomful of batteries, some 2,000 cells in all, to make enough electricity to produce pure potassium metal by the process of electrolysis (pp. 32-33).

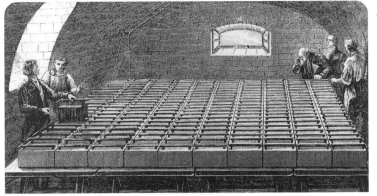

HIGH-TENSION BATTERY
This 1920s lead-acid storage battery was used to provide electricity for a domestic radio set. "High tension" means that the electricity had a high pressure, that is, a large voltage. Each glass jar contained two lead plates and the electrolyte – dilute sulfuric acid. The battery was recharged at the local garage, where batteries used in the new and popular automobiles were maintained, by being connected to an electricity supply to reverse the chemical reaction. It was then topped up with water to compensate for any water loss through evaporation.

High-tension battery without lid

Copper anode

Wire carries electric current

Wooden case

Glass containers – each one a single cell

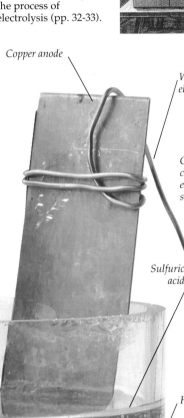

Sulfuric acid

Hydrogen gas

Connecting bars

Swinging bar lifts electrodes up when battery not in use

PORTABLE LIGHT TO READ BY
A safe, portable battery, combined with the electric light bulb (pp. 46-47), supplied light wherever it was needed. This late 19th-century train traveler reads with the battery in the bag by his side.

Sealed top-up hole

Terminal

1940s car battery

THE CAR BATTERY
These rechargeable lead-acid batteries are known as storage batteries. Each cell consists of two lead plates, or electrodes, separated by sulfuric acid. As the battery is charged, lead oxide forms on one of the plates, storing the incoming electrical energy in chemical form.

DANIELL CELL
English professor John Daniell (1790-1845) developed a simple cell in 1836 that provided current for a longer period. His cell (right) had a copper cylinder as the positive electrode (anode) in copper sulfate, and a zinc rod as the negative electrode (cathode) in sulfuric acid, separated by a porous pot. It produced about one volt and supplied electricity for research.

Sulfuric acid fills void

Cell divider

Lead oxide plate

Lead metal plate

Circuits and conductors

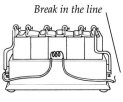

HENRY CAVENDISH
Forgetful and reclusive, Cavendish (1731-1810) was a great experimenter. His electrical research was wide ranging, and like other scientists of the time he assessed the intensity of electricity by how severely he was shocked by it. One of Cavendish's theories was the idea of electric potential or voltage to explain how the "push" of the electricity was produced.

As SCIENTISTS EXPERIMENTED with batteries, they discovered that some substances let electric charge pass through them without difficulty, while others would not. The former were called conductors, and the latter insulators. But why do some substances conduct? It is because all matter is made of atoms (pp. 54-55). An atom, in turn, is made of particles called electrons orbiting around a central nucleus. Each electron has a negative charge, and the nucleus has balancing positive charges. An electric current occurs when the electrons move. Certain substances, especially metals, have electrons that are not tightly held to their nuclei. These "free electrons" are more mobile, and they can be set in one-way motion easily, to produce an electric current.

INSIDE A CONDUCTOR
Metals are good conductors. In a good conductor, each atom has one or more free electrons. The nucleus cannot hold on to them strongly, and they can move within the conductors. Normally this happens at random, and there is no overall one-way flow of electrons. But if there is a difference in voltage between one end of the conductor and the other, the negative electrons are attracted to the positive end and so the current flows.

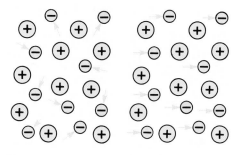

Electrons move at random

Flow of electrons producing current

MAKING A CIRCUIT
A cell or battery on its own does not produce an electric current. Electrons flow only if they have somewhere to go, and if a voltage pushes them. When conductors link one terminal of a battery to the other, a current flows. This series of conductors is a circuit. In an open or broken circuit, there is a break along the line, and the current stops. In a closed or complete circuit (right), the electrons flow from the negative side of the battery to the positive side. They can then do work, such as producing light.

Open circuit

Break in the line

Closed circuit

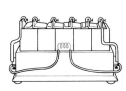

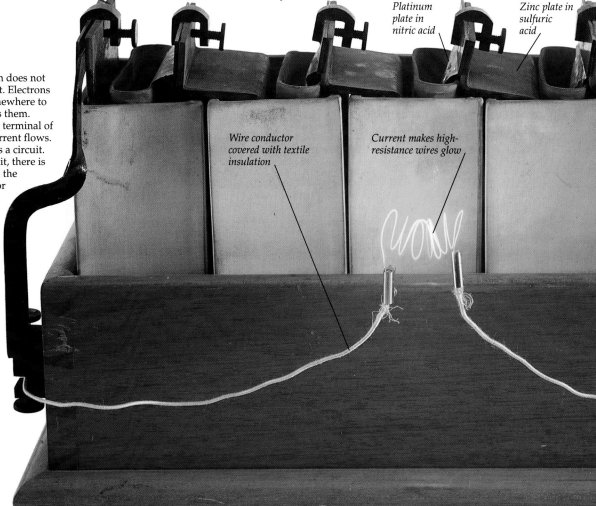

Battery

Platinum plate in nitric acid

Zinc plate in sulfuric acid

Wire conductor covered with textile insulation

Current makes high-resistance wires glow

Carrying current

Early experiments on conductors and insulators showed that most metals were good at carrying current. Soon wires of iron or, even better, copper and silver were being used in electrical research. Some forms of carbon, such as graphite (the "lead" in a pencil) and charcoal were also conductors (p.37). So was any watery substance, from solutions of acids and salts, to parts of vegetables and animals. Insulators tended to be woody, fibrous, or made of minerals. An early list of insulators included leather, parchment, ashes, chalk, hair, porcelain, feathers, precious stones, dried vegetables, resin, and amber.

A good conductor

A poor conductor

Charcoal conducts the charge away rapidly

Wool conducts the charge away very slowly

TESTING CONDUCTORS
An electroscope can show whether a substance was a conductor or an insulator. It was charged from a Leyden jar or electrostatic generator so that its two gold leaves repelled each other and moved apart. Then the test substance was held in the hand, against the metal cap. If the substance was a good conductor, it conducted the charge away from the electroscope, to the body and then into the ground. The leaves would lose their charge and flop back down. The speed with which they came to rest together showed how fast their charge was being conducted away, and so indicated the efficiency of the conductor.

Gold leaves lose their charge and close together

Porous pot

Gold leaves keep their charge and repel each other

INSULATING PLUGS
For safety's sake, the outside of a plug must be well insulated from the conducting metal prongs inside. Bodies of early plugs were made of wood and porcelain. Bakelite, a synthetic resin and a good insulator, came into use in the 1910s.

Bakelite body (1930)

Wooden body (1915)

Porcelain body (1920)

WEATHERPROOF INSULATION
High-voltage power lines must be well insulated or the electricity may force its way through a poor insulator or spark across a gap. Porous materials like wood are not used because they can absorb moisture. Nonporous substances such as ceramics and glass are shaped to keep rain, dew, and other moisture from creating a circuit into the ground.

Early porcelain insulator on a telegraph pole

Ceramic insulator (1956)

Toughened glass suspension insulator (1956)

Aluminum power line

Resisting electricity

IN A SERIES OF EXPERIMENTS around 1825, the German scientist Georg Ohm (1787-1854) demonstrated that there were no perfect electrical conductors. Each type of substance, even the best metals, put up some resistance to the current. Ohm showed that a long wire had more resistance than a short one of the same metal, and thin wires had more resistance than fat ones. Also, in a circuit, the greater the resistance, the more potential difference (volts) was needed to push the current through the wire. The relationships between potential difference, current, and resistance became known as Ohm's law.

ADJUSTABLE RESISTANCE
Scientists needing to vary the current in a circuit developed adjustable resistors, or rheostats. (Their uses today include dimmer switches.) One simple design used special resistance wire, made from a combination of metals, such as nickel and copper. This put up some resistance to the current, but not too much. A long piece of the resistance wire was wound as a coil on an insulating tube. This was more convenient than having it stretched out straight. The wire touched a contact that slid along the top. Electricity from a battery came in through one terminal, then into the resistance coil. As the contact slid one way, the electricity went through more resistance wire, so the rheostat's resistance rose. As the contact moved the other way, its resistance fell.

Ammeter shows larger current

Sliding contact is a short distance along the tube

Rheostat

Wire to battery

Wire connecting rheostat to battery

Ammeter

Insulating tube

1 LOWER RESISTANCE
Electricity passes through only a short length of resistance wire, then through the top bar to the red wire. The circuit has a low overall resistance, so a large current flows, as shown on the meter.

Ammeter shows smaller current

Sliding contact half way along the tube

2 HIGHER RESISTANCE
The electricity now flows through more resistance wire. (The bar and other wires have hardly any resistance.) The circuit has a higher overall resistance, so a smaller current flows, as shown on the meter.

Georg Ohm

Ohm's practical experiments showed the mathematical links between resistance, potential difference, and current. Ohm's law of 1826 states that provided the temperature does not change, the current flowing through certain conductors is proportional to the potential difference (voltage) across it. This is written as I (current) = V (voltage) divided by R (resistance) and is now a cornerstone of electric circuit design.

Coiled resistance wire

Metal bar

Torsion head twists to bring magnetic needle back to zero

Suspension wire

Magnetized needle turned by current in copper bar

Magnifying glass to detect swing of the needle

Cups of mercury into which ends of resistor are dipped

Copper bar

Bismuth bar

Container for ice

Boiler producing steam

OHM'S APPARATUS
To test his theory, Ohm used a thermocouple (below), which produced a small voltage when there was a temperature difference between the junctions of two metals (pp. 52-53). To measure the current through the wire under test, he used a torsion balance similar to the one used by Coulomb (p. 11). In this reconstruction of his apparatus, he measured the deflection of the magnetized needle.

Cold junction of thermocouple

Hot junction of thermocouple

STANDARD RESISTOR
After Ohm's work revealed the importance of resistance, standard resistors like this were made. It was connected up in an electric circuit and measurements were made to calculate the resistance using Ohm's law. In 1861 the British Association Standards Committee was set up to determine the most convenient unit for resistance. It was adopted in 1865. Later, it was named "ohm" with the symbol Ω.

Brass casing for resistance wire

Copper connecting leads

Wooden storage box

THE SEARCH FOR BETTER CONDUCTORS
With a knowledge of resistance, engineers could improve the design of long-distance electrical and communication cables. Better conductors surrounded by better insulators would carry higher currents farther, without expensive loss of electricity. Copper was the preferred conductor. After silver, it has the lowest resistance; it does not rust or corrode like iron or steel; and it is easily drawn out into a wire.

Power cable (20th century)

Copper cores

Flexible lead sheath

Paper insulation

Telegraph cable (1837)

Resin

Wood

Copper cores

Measuring and quantifying

IN THE MIDDLE OF THE 19TH CENTURY scientists were busy studying electricity and its effects. They experimented with new types of cells and batteries, and with circuits containing different components. There was still no widespread commercial use for electricity, except for the telegraph (pp. 56-57). As with all scientific research, the experimenters had to find ways of measuring, recording, and checking their work. New measuring devices were needed and meters of ingenious design were used by the researchers who struggled to understand this mysterious "force." To make measuring devices useful, new units were devised and named. The existing units, such as inches and ounces, were no use for the flowing charge of an electric current.

Wires connected to measured supply of electricity

JOULE'S LAW
An electric current flowing through a high-resistance wire heats the wire (pp. 48-49), which warms a known amount of water. The temperature rise of the water over a certain time, compared to the electric current passed through the wire, shows the relationship between the two. Joule's law states that the heat produced is proportional to the resistance of the wire, the time the current flows, and the square of the current. His work was a further stage in the development of electricity as another branch of science.

Thermometer measures temperature rise

ELECTRICITY AND HEAT
The work of Englishman James Joule (1818-1889) was dominated by his study of the connections between heat, electricity, and mechanical work. Joule showed that electricity and mechanical work did not just move heat from one place to another, as was believed at the time; they generated heat. Heat could also be turned into mechanical work. Joule's work was the basis of the modern concept of energy. A modern unit of energy, the "joule," is named after him.

Coiled resistance wire

Measured volume of water

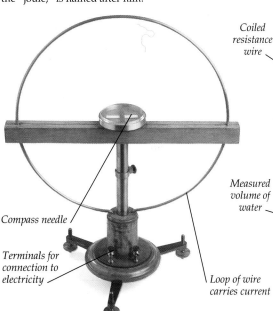

Compass needle

Terminals for connection to electricity

Loop of wire carries current

THE GALVANOMETER
Joule used a tangent galvanometer in his experiments. Current flowing in the loop of wire produces a magnetic field. When the meter is set up correctly, the field deflects a small magnetic needle from its normal north-south alignment with the Earth's magnetic field. The current is proportional to the amount the needle is deflected from its original ("at rest") position.

INVENTION OF THE GALVANOMETER
Leopold Nobili (1784-1835) devised one of the earliest sensitive electrical measuring instruments, the astatic galvanometer (pp. 34-35). The galvanometer indicated current – the quantity of charge moving past a certain point per unit of time. In the 19th century, electric current produced by batteries was called galvanic current after Galvani (pp. 16-17).

AN AMMETER
By 1890 the expanding electricity business brought the need for portable, accurate, robust meters. This moving coil meter showed the current in amps.

Flowing currents

Electricity has been likened to a fluid flowing unseen from place to place. Some of the words used to describe electricity, such as "current" and "flow," relate to these notions. The comparison between electricity passing along wires and water flowing through pipes is not an exact parallel, but it can help to explain some of electricity's stranger properties by making them more physical and familiar.

WATER ANALOGY
The rate of water flow (the volume passing a certain point in a given time) is similar to the current in an electric circuit, measured in amps. The pressure, or pushing force, of the water can be thought of as the potential difference in an electric circuit, measured in volts. A narrower pipe resists water flow, as thin wire in a circuit resists the flow of electricity.

Flow meter shows two units of flow in this part of the circuit, where the water flow is slowed down because it had to turn the propeller

THE PROPELLER
The propeller turns when the water flows by it. It resists the flow of water and this results in a slower flow, and a reduction in the number of units shown on the meter.

Propeller

RESISTANCE TO FLOW
Certain factors set up a resistance to flow (pp. 22-23). The propeller and the coils of tubing do this in the circuit shown here. The flow meters show less flow or current in their branches of the circuit, but the flows add up to the total flow from the pump. In electricity, resistance to the current could be provided by an electric motor or a conducting wire.

Flow meter shows one unit of flow in this part of the circuit, where there is resistance to flow set up by the long, thin pipe

A resistance coil is made up of a long length of thin piping wound into a coil to make it more manageable in the circuit

METERS
Meters show the units of flow in the pipes, and they register how these units are shared between the parallel pipes in the circuit. In electricity these would be ammeters, measuring the current in amps.

POWER SOURCE
The pump represents the battery or other electricity source which provides the "pushing power," or pressure, to force the water through the pipes. In electricity this is measured by the voltage.

Flow meter shows three units of flow as the water is pushed around the circuit by the pump

Pump provides the pushing power

Water flows in a counter-clockwise direction

THE INTERNATIONAL VISUAL LANGUAGE OF ELECTRICITY

It would be time consuming if all the information about electricity was written out in full. Symbols are used as an international visual language that is understood by electrical engineers, circuit designers, and teachers.

A = ampere/amp (p. 27)
AC = alternating current (pp. 40-41)
C = coulomb (p. 11)
DC = direct current (pp. 18-19)
emf = electromotive force (p. 20)
F = farad (p. 13)
H = henry (p. 35)

J = joule (left)
KWH = kilowatt-hour (pp. 60-61)
p.d. = potential difference (p. 22)
V = volt (pp. 20-21)
W = watt (p. 42)
Ω = ohm (pp. 22-23)

AC supply (pp. 40-41)
ammeter (pp. 22-23)
battery (pp. 18-19)
cell (pp. 16-19)
fuse (pp. 46-47)

relay (pp. 30-31)
resistor (pp. 22-23)
switch (pp. 46-49)
transformer (pp. 40-41)
voltmeter (pp. 32-33)

Magnetism from electricity

With the development of Volta's battery in 1800, scientists had a source of steadily flowing electric current. This opened up new fields of research. Twenty years later an observation by Hans Christian Oersted (1777-1851) from Copenhagen began to link the two great scientific mysteries of the age: electricity and magnetism. Oersted noticed that a metal wire carrying a current affected a magnetic compass needle that was brought near to it. This revelation, which contradicted the orthodox philosophy of the time, was published in scientific journals. French scientific thinker André-Marie Ampère (1775-1836) heard about Oersted's work from a fellow scientist. He doubted it at first, so he repeated the tests, but had similar results. Ampère therefore set to work to describe the effect more fully and to explain the connection between electricity and magnetism. In repeating the experiments, he provided a theoretical and mathematical description of the practical results of Oersted's work. From this flowed discoveries such as the electromagnet (pp. 26-27) and the telegraph receiver (pp. 56-57).

OERSTED'S ANNOUNCEMENT
Oersted published his discovery of the interaction between electricity and magnetism on July 21, 1820. The pamphlet was written in Latin but was translated into various languages including his own, Danish. Michael Faraday (pp. 34-35) would probably have heard about the discovery in this way.

Current off

With no current in the wire, the filings lie haphazardly

THE MAGNETIC FIELD
Magnetism, like electricity, is invisible – but its effects can be seen. Iron-containing substances such as iron filings are attracted to an ordinary bar magnet, and they line up to indicate the direction of the invisible "lines of force" of the magnetic field. An electric current also creates a magnetic field. With no current in the wire, the filings lie randomly on the card. Switch the current on, tap the card, and the filings line up to reveal a circular magnetic field around the wire.

Current on

Filings line up indicating circular magnetic field

Wooden clamp

Current-carrying wire

Dilute acid between plates

Battery

Battery terminal

AMPERE'S ACHIEVEMENT

Ampère developed the science of electrodynamics. He proved that the strength of the magnetic field around a wire, shown by the amount of the compass needle's deflection, rises with increasing current, and decreases with the distance away from the wire. Ampère extended Oersted's work, but he could not see a clear relationship between the needle's movements and the position of the wire. Then he realized that the direction in which the needle settled depended on the earth's magnetic field as well as the magnetic field produced by the current. He devised a way of neutralizing the earth's magnetic field and found that the needle then settled in the direction of the magnetic field produced by the current and not with the earth's north-south magnetic field. Ampère also found that two parallel electric currents had an effect on each other. If the currents run in the same direction, they attract each other; if they run in opposite directions, they repel each other. To commemorate Ampère's achievements, the modern unit of current – the ampere (shortened to "amp") – was named after him.

André-Marie Ampère

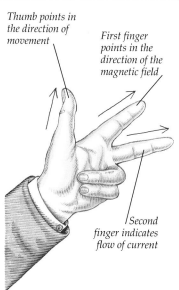

Thumb points in the direction of movement

First finger points in the direction of the magnetic field

Second finger indicates flow of current

THE LEFT-HAND RULE

When an electric current crosses a magnetic field, then the magnetic field, the current, and the force on the current – and thus the movement of the wire carrying the current, if it can move (pp. 38-39) – are all in different directions. A current-carrying wire or other conductor in a magnetic field tries to move according to this handy rule: With the thumb and first two fingers of the left hand at right angles to each other, the first finger shows the direction of the magnetic field (points from the north pole to the south pole of the magnet); the second finger shows the flow of the electric current (points from positive to negative); and the thumb shows the movement of the wire.

Reconstruction of Oersted's experiment

Needle deflected from north-south by magnetic field of current in wire

Wooden clamp

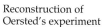
South

North

Current-carrying wire

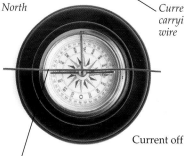

Current on

Current off

Needle aligns with earth's north-south magnetic field

Hans Christian Oersted

THE PROCESS OF DEDUCTION

During a lecture in Copenhagen on changing electricity to heat, Oersted moved a compass near the current-carrying wire. The wire caused the compass needle – which is a magnet – to swing away from its normal north-south alignment with the earth's magnetic field. In this reconstruction of his demonstration, the stands and the bench are made from an insulating material so as not to disturb the magnetic field. Oersted had shown that magnetism had been produced by the electric current and this phenomenon is known as electromagnetism. Oersted's discovery became the basis of the electric motor (pp. 38-39) and the electromagnet (pp. 28-29) which is involved in so many areas of our daily lives from the telephone (pp. 58-59) to the starting of a car (pp. 30-31).

Plates of copper and zinc soldered back-to-back and cemented into wooden trough

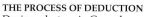
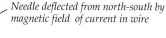

Electromagnets

T HE DISCOVERY OF a magnetic field around a current-carrying wire (pp. 26-27), and the fact that a coil of wire had a greater magnetic effect than a single turn, led to a fascinating new gadget that made lecture audiences gasp with surprise. In 1825 William Sturgeon (1783-1850) wound a coil of wire around an iron rod and built one of the first electromagnets. An electromagnet differs from the usual permanent magnet – its magnetism can be turned on and off. It usually consists of insulated electrical wire wound around a piece of iron, known as the core. Switch on the current, and the magnetic field around the wire makes the core behave like a magnet. It attracts iron-containing substances in the usual way. Switch off the electricity, and the magnetism disappears. Sturgeon built the electromagnets as demonstration models for his lectures.

ELECTROMAGNETIC CHAIR
A chair was a convenient base for this great horseshoe electromagnet of Michael Faraday (pp. 34-35)). It was used to investigate the effects of magnetism.

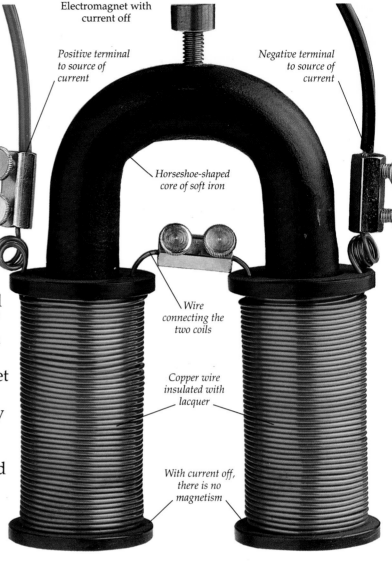

Electromagnet with current off

Positive terminal to source of current

Negative terminal to source of current

Horseshoe-shaped core of soft iron

Wire connecting the two coils

Copper wire insulated with lacquer

With current off, there is no magnetism

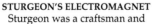

Fabric insulation on iron core

Coil of wire

Sturgeon's electromagnet

Iron core

End connected to battery

End connected to battery

STURGEON'S ELECTROMAGNET
Sturgeon was a craftsman and instrument maker who published catalogues of his electromagnetic apparatus (left). He enjoyed the practical "how-to" side of science, but had less concern for theories about why things happened. His electromagnet (above) depended on current passing through a coil of wire, to create an appreciable magnetic field. In his time, wires were made bare and uninsulated. (Later electromagnets had insulated wire.) He insulated the iron core to stop the electricity passing straight along it instead. The turns of the coil were also spaced apart, to prevent the current jumping straight from one to the next where they touched – short circuiting.

INVISIBLE LIFTING POWER
This modern electromagnet demonstrates its strength by attracting a loose pile of iron filings when the current is switched on. In a bar-shaped electromagnet, one end of the core becomes the north pole and the other is the south pole, depending on the current's direction of flow. This horseshoe-shaped core has two coils for a stronger magnetic field between its ends (bringing the poles closer together makes the magnet more powerful). Sturgeon made an initial discovery that soft iron was a good metal for the core, since it became magnetized more easily than steel. In the early days, there were races to build the biggest electromagnets and lift the heaviest weights. By the late 1820s electromagnets in Europe could lift around 11 lb (5 kg). In the United States, Joseph Henry realized that more turns in the coil produced a stronger magnetic field (up to a limit). It is said that he made insulated wire by wrapping bare wire with strips of silk from his wife's clothes. He could then pack more turns of wire into a smaller space, without the risk of short-circuit. The greater the current, the more powerful was the magnetic field. Henry's large electromagnet (above right) lifted about 750 lb (340 kg).

Iron filings

Electromagnet with current on

Positive terminal to source of current

Negative terminal to source of current

Horseshoe-shaped core of soft iron

Wire connecting the two coils

Copper wire insulated with lacquer

Force of magnetic field overcomes gravity and lifts iron filings

Henry's electromagnet

Nonmagnetic stand

Iron core

Several layers of insulated wire

Zinc plate

Copper plate

Wooden base

JOSEPH HENRY
Henry (1797-1878) was an engineer whose inventive genius led him to improve electromagnets, for example by wrapping a second coil of wire around the first. In this small electromagnet (left), the two metal plates of copper and zinc were immersed in a jar of dilute acid. This formed a voltaic cell (pp. 16-17) that produced the electric current. The wire was wrapped in insulating fabric, so that the coil's turns could be close together for a stronger magnetic field. Henry also developed an early form of telegraph in 1831, although others made fame and fortune from it (pp. 56-57), and some of his work with electromagnetic induction paralleled that of Faraday.

HENRY'S LIFTING RIG
This type of apparatus was used to measure the lifting power of electromagnets made to different designs, from various metals, and with different numbers and arrangements of windings.

ELECTROMAGNETS AT WORK
The lifting power of modern electromagnets is used to separate iron-containing, or ferrous, metals from other materials at the junkyard before recycling.

Electromagnets at work

Since the time of Sturgeon and Henry, electromagnets have been central components in electrical machines. Their power to attract and hold iron-containing materials, for example, when sorting out steel cans from aluminum ones at the recycling center, is not their only use. An electromagnet is a convenient way of turning electrical energy into rotary motion, using the forces of magnetic attraction and repulsion in electric motors (pp. 38-39). The reverse occurs in generators (pp. 36-37). Electricity is converted into push-pull movements by electromagnetic relays and solenoids. The movements operate mechanical devices, such as telegraph receivers. Electro-magnetism is also used in electrical transformers (pp. 40-41), and to manipulate atoms in particle accelerators.

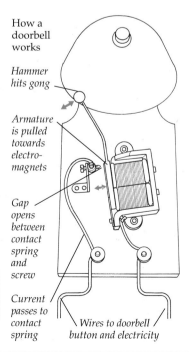

SAVING SIGHT
A finely controlled electromagnet could remove iron-containing foreign particles from the tissues of the eye. It pulled fragments straight out, minimizing the risk of further damage.

RINGING A BELL
Pressing the button of this doorbell completes a circuit in which the current passes into one terminal of the doorbell, then through the coils of the electromagnets, along the springy steel contact, into the brass screw and its mounting, and out through the other terminal. As the armature is pulled towards the electromagnets, the hammer hits the gong. At the same time, a gap opens between the contact spring and the screw, which breaks the circuit and switches off the electricity. The magnetism disappears, the spring on which the armature is mounted moves the armature back into its original position, the contacts touch and make the circuit again. The process is repeated until the button is released.

Gong

Hammer

Contact points

Contact-adjusting brass screw

Steel contact spring

Soft iron armature

Electromagnet

Spring mounting for armature

How a doorbell works

Hammer hits gong

Armature is pulled towards electromagnets

Gap opens between contact spring and screw

Current passes to contact spring

Wires to doorbell button and electricity

THE DOORBELL
When the current is switched on, the electromagnet attracts a nearby piece of iron. The movement of the iron, known as the armature, can be used to switch on a separate electrical circuit. In this doorbell the armature also switches off its own circuit. Huge rows of relays are still the basis for some telephone exchanges (pp. 58-59).

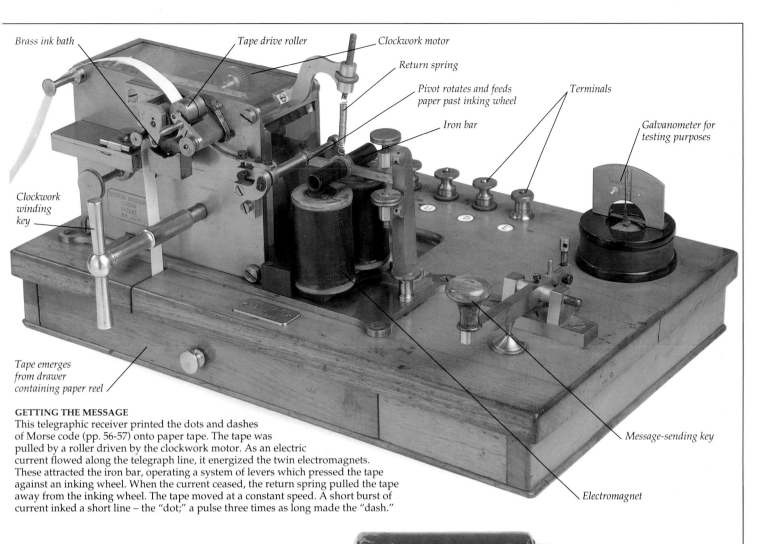

Brass ink bath

Tape drive roller

Clockwork motor

Return spring

Pivot rotates and feeds
paper past inking wheel

Iron bar

Terminals

Galvanometer for
testing purposes

Clockwork
winding
key

Tape emerges
from drawer
containing paper reel

Message-sending key

Electromagnet

GETTING THE MESSAGE

This telegraphic receiver printed the dots and dashes
of Morse code (pp. 56-57) onto paper tape. The tape was
pulled by a roller driven by the clockwork motor. As an electric
current flowed along the telegraph line, it energized the twin electromagnets.
These attracted the iron bar, operating a system of levers which pressed the tape
against an inking wheel. When the current ceased, the return spring pulled the tape
away from the inking wheel. The tape moved at a constant speed. A short burst of
current inked a short line – the "dot;" a pulse three times as long made the "dash."

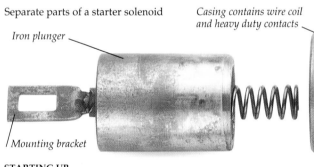

Separate parts of a starter solenoid

Iron plunger

Casing contains wire coil
and heavy duty contacts

Heavy duty
wires to
starter motor

Mounting bracket

STARTING UP

A solenoid is a cylindrical coil of wire, usually with
an iron core or plunger inside. At rest, the plunger
sticks out of the coil. When the current is switched
on, the plunger is attracted into the magnetic field
created inside the coil, and pulled strongly into the
middle of the coil. In a car the key-operated
ignition switch completes a circuit and
energizes the starter solenoid. The
iron plunger moves and closes
a switch that completes a
separate circuit through
the starter motor, which
turns the engine. This
separation is necessary
because a starter
motor draws a high
current from the car's
battery which would
otherwise deliver a
shock to the driver.

Ignition key
switches on
small current

Low-current circuit

Plunger

Car battery

Heavy
duty contacts

Motor is
activated

Starter
solenoid is
energized

Starter motor receives high current

ATOMIC PARTICLES

The study of sub-
atomic particles
involves accelerating
them to incredible
speeds (pp. 12-13).
A magnetic field
deflects charged
particles, so powerful
electromagnets can
be adjusted to control,
bend, and focus their
paths. These long,
square electromagnets
on the left are at a
nuclear research
center near Geneva
in Switzerland.

31

Discoveries using electricity

IN 1800 A FEW WEEKS AFTER VOLTA announced his electricity-making pile (pp. 16-17), William Nicholson (1753-1815) and Anthony Carlisle (1768-1840) were experimenting with their own version. They noticed bubbles appearing in a drop of water on top of the pile. They then passed the electric current through a bowl of water. Bubbles of gas appeared around the two metal contacts where they dipped into the water. The bubbles around one contact were hydrogen, and the others were oxygen. The electric current had produced a chemical reaction, splitting water into its two elements. This was electrolysis. Soon other substances were being investigated. Humphry Davy brought electrolysis to prominence when he used it to make pure potassium and sodium for the first time. Today's uses include electroplating, refining pure copper, extracting aluminum from its treated natural ores, and making substances such as chlorine.

HUMPHRY DAVY
In 1807 Davy (1778-1829) made pure potassium by electrolysis. With a bigger electric current (p. 19), he made pure sodium. He helped popularize science, demonstrating such wonders as the intense light of an arc lamp (p. 37).

VOLTAMETER
This modern demonstration apparatus shows how gases are produced by electrolysis. A replica of the original glass apparatus is shown on p. 2. Faraday named it a "volta electrometer" because the gas collected was proportional to the quantity of electricity made by a voltaic pile (pp. 16-17).

Rising bubbles of hydrogen gas

Negative electrode (cathode)

Connecting wire

Watertight rubber plug

To electricity supply

Battery as source of current

Current-carrying wires

Potash melted by current

MAKING PURE POTASSIUM FOR THE FIRST TIME
Pure potassium is a soft, silvery-white metal. It does not occur on its own in nature, because it combines so readily with other substances to form compounds. One is potash (potassium carbonate), found in the ashes of burned plants. This is an ingredient in certain types of glass. Davy passed a large current through molten potash in a metal pot, and pure potassium collected around the negative contact. He named the new substance potassium after its source. In the same way he prepared sodium from soda (sodium carbonate). Davy's successor at the Royal Institution, Michael Faraday, also studied electrolysis. Two laws of electrolysis are named after him.

Potassium reacts with air so must be sealed

Pure potassium in a sealed glass tube

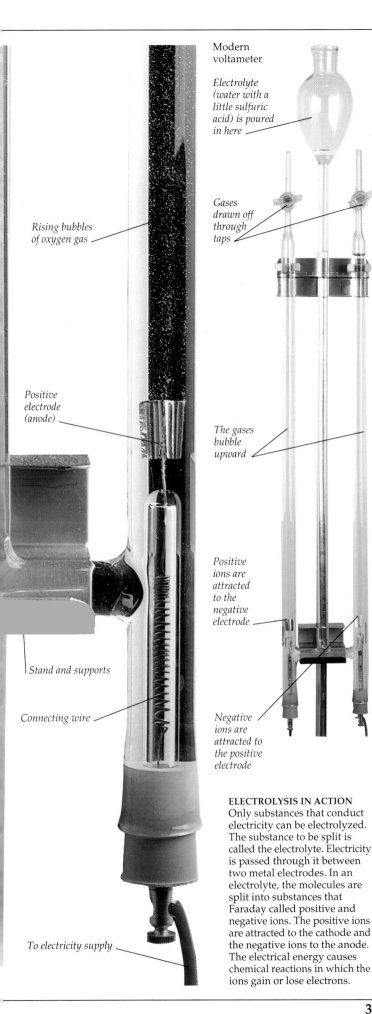

Modern voltameter

Electrolyte (water with a little sulfuric acid) is poured in here

Rising bubbles of oxygen gas

Gases drawn off through taps

Positive electrode (anode)

The gases bubble upward

Stand and supports

Positive ions are attracted to the negative electrode

Connecting wire

Negative ions are attracted to the positive electrode

To electricity supply

ELECTROLYSIS IN ACTION
Only substances that conduct electricity can be electrolyzed. The substance to be split is called the electrolyte. Electricity is passed through it between two metal electrodes. In an electrolyte, the molecules are split into substances that Faraday called positive and negative ions. The positive ions are attracted to the cathode and the negative ions to the anode. The electrical energy causes chemical reactions in which the ions gain or lose electrons.

Electroplating

In this process, an object receives a coating of a metal by electrolysis. The object to be plated is connected to the negative terminal of a battery or similar electricity supply so that it acts as the negative electrode (cathode). It is left for a time in a solution containing a compound of the metal that will form the plating. For example, an object to be plated with copper is put into copper sulfate solution. This solution is the electrolyte. In it, the copper exists as positively charged ions. When the electric current flows, these ions are attracted to the negative electrode – the object. They settle evenly on its surface.

Nickel spoon

Silver plated spoon

ELECTROPLATING A SILVER SPOON
Solid silver cutlery is expensive, so cutlery is often made from a cheaper metal and thinly plated with silver by electrolysis. The item must be clean. Otherwise, the silver will not adhere to its surface. Silver is transferred from the solution to the object.

Electroplating tank

Electricity supply

Spoon cathode slowly revolves to ensure even plating

Silver nitrate solution

Silver anode

RUST-RESISTANT CARS
Steel vehicle bodies can be electroplated with a thin layer of a metal such as zinc to protect them from rust. The car body (the cathode) is electrified, and the zinc (the anode) is drawn from the solution to cover every nook and cranny.

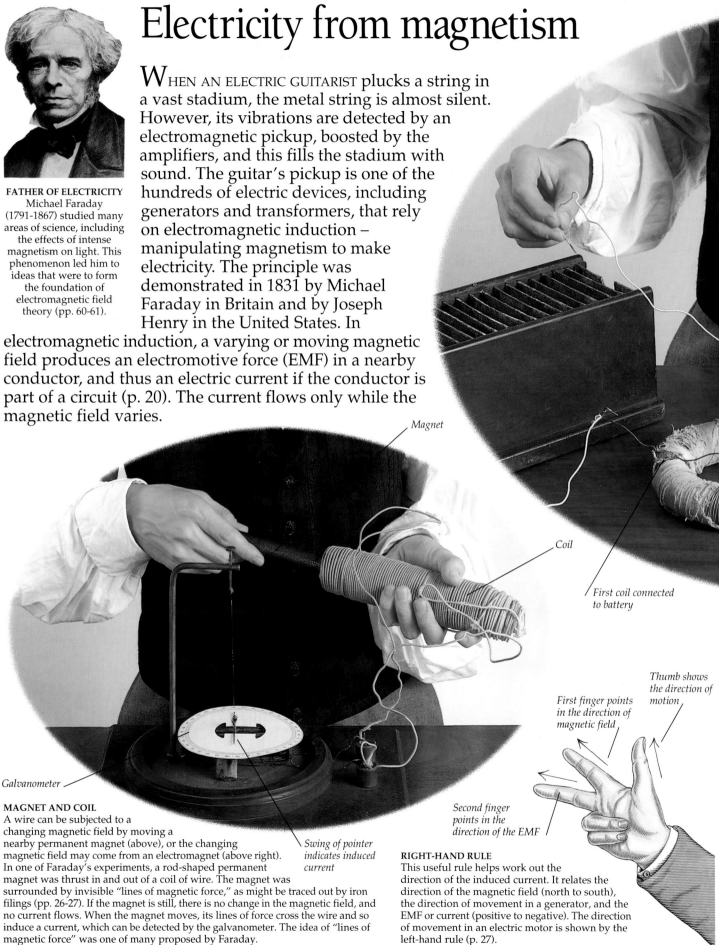

Electricity from magnetism

W HEN AN ELECTRIC GUITARIST plucks a string in
a vast stadium, the metal string is almost silent.
However, its vibrations are detected by an
electromagnetic pickup, boosted by the
amplifiers, and this fills the stadium with
sound. The guitar's pickup is one of the
hundreds of electric devices, including
generators and transformers, that rely
on electromagnetic induction –
manipulating magnetism to make
electricity. The principle was
demonstrated in 1831 by Michael
Faraday in Britain and by Joseph
Henry in the United States. In
electromagnetic induction, a varying or moving magnetic
field produces an electromotive force (EMF) in a nearby
conductor, and thus an electric current if the conductor is
part of a circuit (p. 20). The current flows only while the
magnetic field varies.

Magnet

Coil

*First coil connected
to battery*

*Thumb shows
the direction of
motion*

*First finger points
in the direction of
magnetic field*

*Second finger
points in the
direction of the EMF*

Galvanometer

*Swing of pointer
indicates induced
current*

MAGNET AND COIL
A wire can be subjected to a
changing magnetic field by moving a
nearby permanent magnet (above), or the changing
magnetic field may come from an electromagnet (above right).
In one of Faraday's experiments, a rod-shaped permanent
magnet was thrust in and out of a coil of wire. The magnet was
surrounded by invisible "lines of magnetic force," as might be traced out by iron
filings (pp. 26-27). If the magnet is still, there is no change in the magnetic field, and
no current flows. When the magnet moves, its lines of force cross the wire and so
induce a current, which can be detected by the galvanometer. The idea of "lines of
magnetic force" was one of many proposed by Faraday.

RIGHT-HAND RULE
This useful rule helps work out the
direction of the induced current. It relates the
direction of the magnetic field (north to south),
the direction of movement in a generator, and the
EMF or current (positive to negative). The direction
of movement in an electric motor is shown by the
left-hand rule (p. 27).

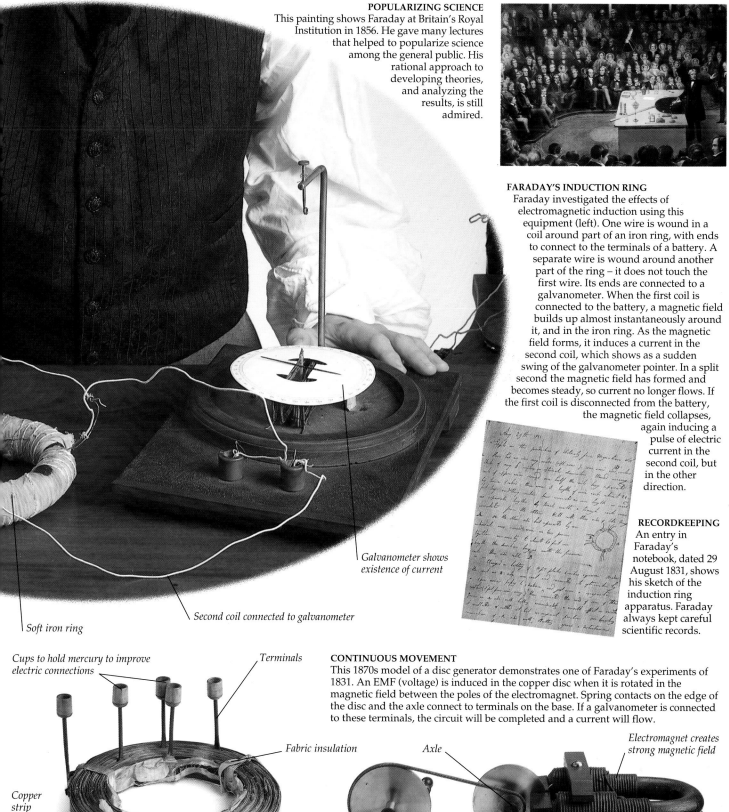

POPULARIZING SCIENCE
This painting shows Faraday at Britain's Royal
Institution in 1856. He gave many lectures
that helped to popularize science
among the general public. His
rational approach to
developing theories,
and analyzing the
results, is still
admired.

FARADAY'S INDUCTION RING
Faraday investigated the effects of
electromagnetic induction using this
equipment (left). One wire is wound in a
coil around part of an iron ring, with ends
to connect to the terminals of a battery. A
separate wire is wound around another
part of the ring – it does not touch the
first wire. Its ends are connected to a
galvanometer. When the first coil is
connected to the battery, a magnetic field
builds up almost instantaneously around
it, and in the iron ring. As the magnetic
field forms, it induces a current in the
second coil, which shows as a sudden
swing of the galvanometer pointer. In a split
second the magnetic field has formed and
becomes steady, so current no longer flows. If
the first coil is disconnected from the battery,
the magnetic field collapses,
again inducing a
pulse of electric
current in the
second coil, but
in the other
direction.

RECORDKEEPING
An entry in
Faraday's
notebook, dated 29
August 1831, shows
his sketch of the
induction ring
apparatus. Faraday
always kept careful
scientific records.

*Galvanometer shows
existence of current*

Second coil connected to galvanometer

Soft iron ring

*Cups to hold mercury to improve
electric connections*

Terminals

Fabric insulation

Axle

*Electromagnet creates
strong magnetic field*

*Copper
strip
bent into
spiral
shape*

CONTINUOUS MOVEMENT
This 1870s model of a disc generator demonstrates one of Faraday's experiments of
1831. An EMF (voltage) is induced in the copper disc when it is rotated in the
magnetic field between the poles of the electromagnet. Spring contacts on the edge of
the disc and the axle connect to terminals on the base. If a galvanometer is connected
to these terminals, the circuit will be completed and a current will flow.

HENRY'S INDUCTION COIL
Using a coil of copper strips, Joseph Henry recognized another effect
known as self-inductance. He discovered that a wire carrying a
changing current not only induces a voltage in nearby conductors,
but also in itself. This makes it impossible to start and stop a current
instantaneously. The unit of electrical inductance is called the henry.

*Hand crank
turns the disc*

*Copper disc turns in
the magnetic field*

Terminal

*Spring
contact*

Magneto-electric machines

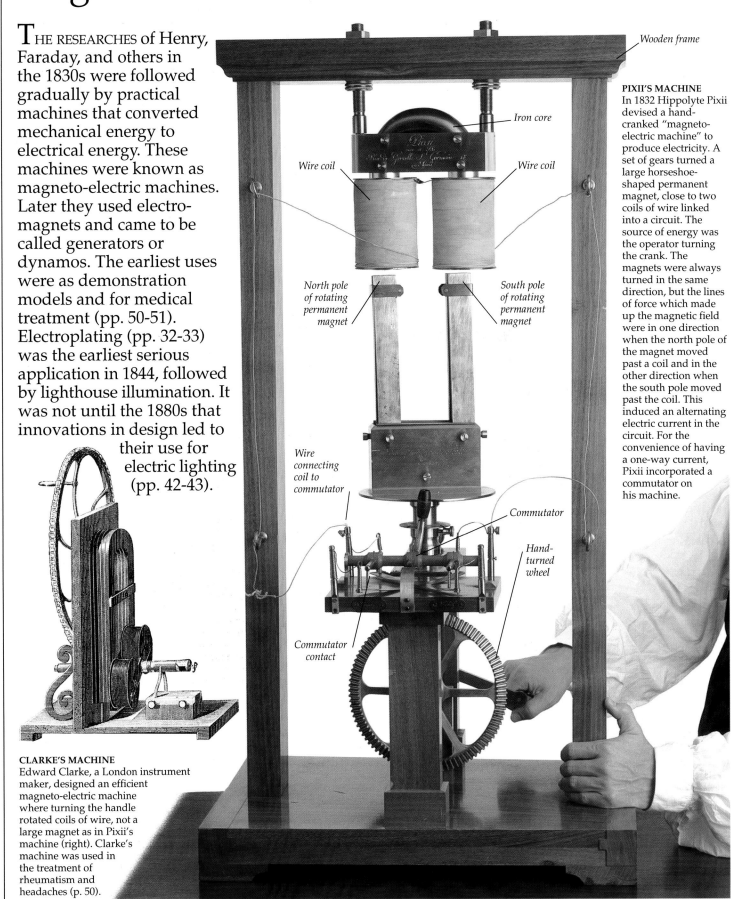

THE RESEARCHES of Henry, Faraday, and others in the 1830s were followed gradually by practical machines that converted mechanical energy to electrical energy. These machines were known as magneto-electric machines. Later they used electro-magnets and came to be called generators or dynamos. The earliest uses were as demonstration models and for medical treatment (pp. 50-51). Electroplating (pp. 32-33) was the earliest serious application in 1844, followed by lighthouse illumination. It was not until the 1880s that innovations in design led to their use for electric lighting (pp. 42-43).

Wooden frame

Iron core

Wire coil

Wire coil

North pole of rotating permanent magnet

South pole of rotating permanent magnet

Wire connecting coil to commutator

Commutator

Hand-turned wheel

Commutator contact

PIXII'S MACHINE
In 1832 Hippolyte Pixii devised a hand-cranked "magneto-electric machine" to produce electricity. A set of gears turned a large horseshoe-shaped permanent magnet, close to two coils of wire linked into a circuit. The source of energy was the operator turning the crank. The magnets were always turned in the same direction, but the lines of force which made up the magnetic field were in one direction when the north pole of the magnet moved past a coil and in the other direction when the south pole moved past the coil. This induced an alternating electric current in the circuit. For the convenience of having a one-way current, Pixii incorporated a commutator on his machine.

CLARKE'S MACHINE
Edward Clarke, a London instrument maker, designed an efficient magneto-electric machine where turning the handle rotated coils of wire, not a large magnet as in Pixii's machine (right). Clarke's machine was used in the treatment of rheumatism and headaches (p. 50).

Coil of
electromagnet

Coil of
electromagnet

Wire gauze
brush

Magnet pole

Commutator

Armature

Drive pulley

Wire gauze
brush

Magnet pole

GRAMME DYNAMO
In about 1870 Belgian inventor Z. T. Gramme devised a dynamo that generated enough electric current, flowing sufficiently steadily, to be useful for large-scale applications such as illuminating factories by carbon arc lights (below). The rotating part of the dynamo, bearing the coils of wire, is known as the armature. Gramme's dynamos were steam-driven and, unlike their predecessors, did not overheat in continuous operation. The Gramme dynamo produced a one-way, or direct, current (DC). It incorporated a commutator – segments of copper which rotated with the armature and caused the current to flow in one direction only. Gauze brushes rubbing against the commutator picked up the current. The dynamo was expensive to maintain because the commutator and brush contacts became worn by pressure and the sparks that flew between them.

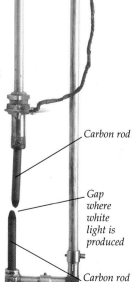

Arc lamp

Electro–
magnetic
mechanism

Carbon rod

Gap
where
white
light is
produced

Carbon rod

WILLIAM SIEMENS
The German Siemens family made many contributions to electrical engineering. William Siemens (1823-1883) designed a dynamo (1866) which used opposing electromagnets to produce a magnetic field around the armature, rather than around a permanent magnet (p. 36).

Carbon rods for lighting
An early use for dynamo-generated electricity was arc lighting. The arc is a "continuous spark" between carbon rods – carbon being a good conductor (p. 21). Arc lamps threw intense light, but they were inefficient, unpredictable, and needed constant attention. Even so, they were installed in public buildings, lighthouses, and used for street lighting.

PUBLIC LIGHTING
In an arc lamp, the gap between the carbon rods is critical. It needs regular adjustment because the gap widens as the carbon vaporizes. A mechanism operated by electromagnets controls the upper carbon rod to keep the necessary gap.

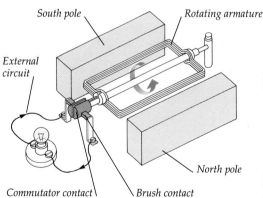

South pole

Rotating armature

External
circuit

Commutator contact

Brush contact

North pole

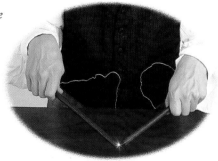

HOW THE DYNAMO WORKS
A coil of wire (the armature) is rotated between the poles of permanent magnets. As one side of the coil travels past the north pole, it cuts the lines of magnetic force, and a current is induced. The coil moves on and the current dies away. Then it approaches the south pole, inducing a current in the opposite direction. The coil is attached to the commutator, which causes the current on the external circuit to flow in the same direction all the time.

HOW THE ARC WORKS
The electric current passes through the pointed tips of the rods as they touch. It heats them so much that the carbon vaporizes. As the rods are drawn apart, the vapor will carry the current across the small gap, glowing intensely as it does so. Davy and Faraday demonstrated arcs at their lectures (pp. 34-35).

19th-century lighthouses at La Hève, France, which used arc lamps

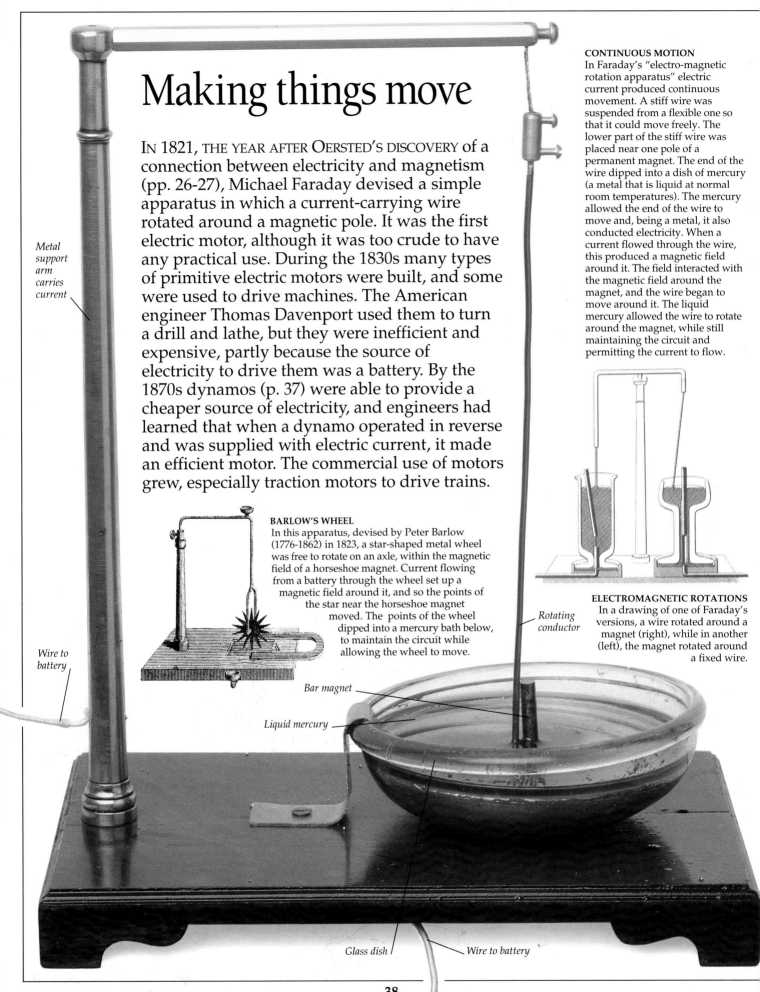

Making things move

IN 1821, THE YEAR AFTER OERSTED'S DISCOVERY of a connection between electricity and magnetism (pp. 26-27), Michael Faraday devised a simple apparatus in which a current-carrying wire rotated around a magnetic pole. It was the first electric motor, although it was too crude to have any practical use. During the 1830s many types of primitive electric motors were built, and some were used to drive machines. The American engineer Thomas Davenport used them to turn a drill and lathe, but they were inefficient and expensive, partly because the source of electricity to drive them was a battery. By the 1870s dynamos (p. 37) were able to provide a cheaper source of electricity, and engineers had learned that when a dynamo operated in reverse and was supplied with electric current, it made an efficient motor. The commercial use of motors grew, especially traction motors to drive trains.

Metal support arm carries current

Wire to battery

CONTINUOUS MOTION
In Faraday's "electro-magnetic rotation apparatus" electric current produced continuous movement. A stiff wire was suspended from a flexible one so that it could move freely. The lower part of the stiff wire was placed near one pole of a permanent magnet. The end of the wire dipped into a dish of mercury (a metal that is liquid at normal room temperatures). The mercury allowed the end of the wire to move and, being a metal, it also conducted electricity. When a current flowed through the wire, this produced a magnetic field around it. The field interacted with the magnetic field around the magnet, and the wire began to move around it. The liquid mercury allowed the wire to rotate around the magnet, while still maintaining the circuit and permitting the current to flow.

BARLOW'S WHEEL
In this apparatus, devised by Peter Barlow (1776-1862) in 1823, a star-shaped metal wheel was free to rotate on an axle, within the magnetic field of a horseshoe magnet. Current flowing from a battery through the wheel set up a magnetic field around it, and so the points of the star near the horseshoe magnet moved. The points of the wheel dipped into a mercury bath below, to maintain the circuit while allowing the wheel to move.

ELECTROMAGNETIC ROTATIONS
In a drawing of one of Faraday's versions, a wire rotated around a magnet (right), while in another (left), the magnet rotated around a fixed wire.

Rotating conductor

Bar magnet

Liquid mercury

Glass dish

Wire to battery

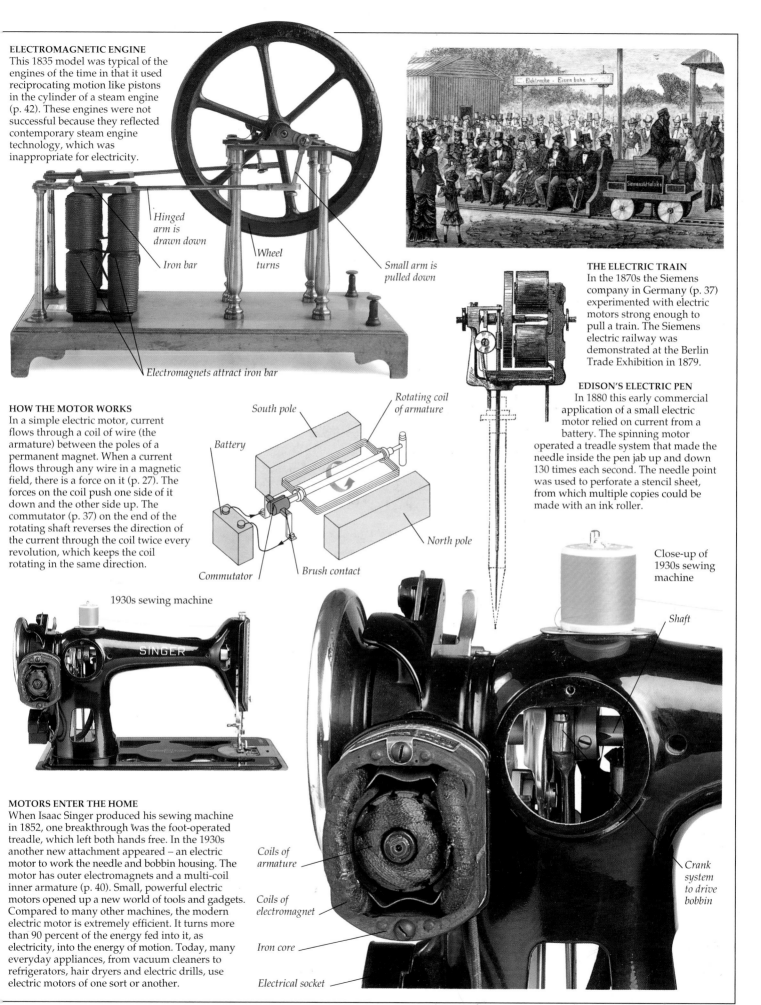

ELECTROMAGNETIC ENGINE

This 1835 model was typical of the engines of the time in that it used reciprocating motion like pistons in the cylinder of a steam engine (p. 42). These engines were not successful because they reflected contemporary steam engine technology, which was inappropriate for electricity.

Hinged arm is drawn down

Iron bar

Wheel turns

Small arm is pulled down

Electromagnets attract iron bar

THE ELECTRIC TRAIN

In the 1870s the Siemens company in Germany (p. 37) experimented with electric motors strong enough to pull a train. The Siemens electric railway was demonstrated at the Berlin Trade Exhibition in 1879.

EDISON'S ELECTRIC PEN

In 1880 this early commercial application of a small electric motor relied on current from a battery. The spinning motor operated a treadle system that made the needle inside the pen jab up and down 130 times each second. The needle point was used to perforate a stencil sheet, from which multiple copies could be made with an ink roller.

HOW THE MOTOR WORKS

In a simple electric motor, current flows through a coil of wire (the armature) between the poles of a permanent magnet. When a current flows through any wire in a magnetic field, there is a force on it (p. 27). The forces on the coil push one side of it down and the other side up. The commutator (p. 37) on the end of the rotating shaft reverses the direction of the current through the coil twice every revolution, which keeps the coil rotating in the same direction.

South pole

Rotating coil of armature

Battery

Commutator

Brush contact

North pole

Close-up of 1930s sewing machine

Shaft

1930s sewing machine

SINGER

MOTORS ENTER THE HOME

When Isaac Singer produced his sewing machine in 1852, one breakthrough Was the foot-operated treadle, which left both hands free. In the 1930s another new attachment appeared – an electric motor to work the needle and bobbin housing. The motor has outer electromagnets and a multi-coil inner armature (p. 40). Small, powerful electric motors opened up a new world of tools and gadgets. Compared to many other machines, the modern electric motor is extremely efficient. It turns more than 90 percent of the energy fed into it, as electricity, into the energy of motion. Today, many everyday appliances, from vacuum cleaners to refrigerators, hair dryers and electric drills, use electric motors of one sort or another.

Coils of armature

Coils of electromagnet

Iron core

Electrical socket

Crank system to drive bobbin

Manipulating electricity

Mᴄʜᴀᴇʟ Fᴀʀᴀᴅᴀʏ's ɪɴᴅᴜᴄᴛɪᴏɴ ʀɪɴɢ (p. 35), with two electrically separate coils, was in effect the first transformer. A transformer can alter the voltage of an electricity supply by having a different number of turns in each coil. Sending electricity long distances through power cables is more efficient at high voltages than at low ones. Transformers boost voltage for transmission, then reduce it at the other end for everyday use. As a transformer works by electromagnetic induction and requires a varying magnetic field, it cannot work with direct current (DC). It operates using alternating current (AC) which rises to a peak flowing one way, fades, and rises again in the reverse direction.

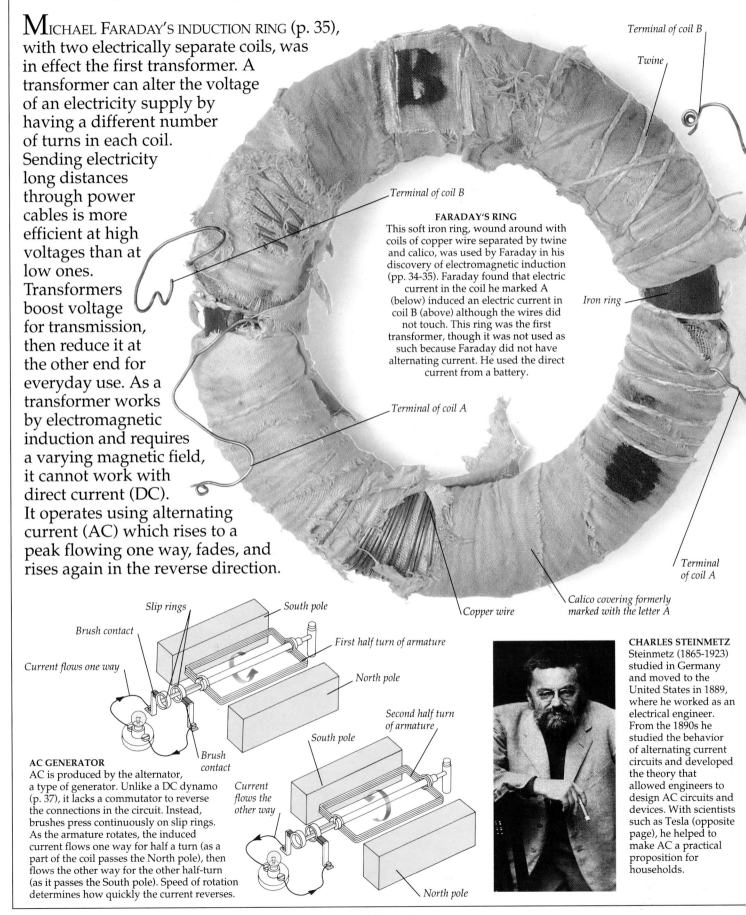

Terminal of coil B

Twine

Terminal of coil B

FARADAY'S RING
This soft iron ring, wound around with coils of copper wire separated by twine and calico, was used by Faraday in his discovery of electromagnetic induction (pp. 34-35). Faraday found that electric current in the coil he marked A (below) induced an electric current in coil B (above) although the wires did not touch. This ring was the first transformer, though it was not used as such because Faraday did not have alternating current. He used the direct current from a battery.

Iron ring

Terminal of coil A

Terminal of coil A

Calico covering formerly marked with the letter A

Copper wire

Slip rings

Brush contact

Current flows one way

South pole

First half turn of armature

North pole

Second half turn of armature

South pole

Brush contact

North pole

Current flows the other way

AC GENERATOR
AC is produced by the alternator, a type of generator. Unlike a DC dynamo (p. 37), it lacks a commutator to reverse the connections in the circuit. Instead, brushes press continuously on slip rings. As the armature rotates, the induced current flows one way for half a turn (as a part of the coil passes the North pole), then flows the other way for the other half-turn (as it passes the South pole). Speed of rotation determines how quickly the current reverses.

CHARLES STEINMETZ
Steinmetz (1865-1923) studied in Germany and moved to the United States in 1889, where he worked as an electrical engineer. From the 1890s he studied the behavior of alternating current circuits and developed the theory that allowed engineers to design AC circuits and devices. With scientists such as Tesla (opposite page), he helped to make AC a practical proposition for households.

Nikola Tesla

Tesla (1856-1943), who was born in Croatia of Serbian parents, emigrated to the United States in 1884. He worked briefly for Thomas Edison. Several years earlier Tesla had realized how he could build an AC motor that would not need a commutator (p. 37). In 1888 he built his first "induction motor." This was a major factor in the widespread adoption of AC supplies. The induction motor is probably the most widely used type of electric motor. Tesla also invented a type of transformer, the Tesla coil, which works at very high frequencies and produces enormous voltages.

READING IN A THUNDERSTORM
Tesla imagined that one day even the power of lightning bolts could be harnessed. This photograph shows him reading at his high-voltage research laboratory in the United States in 1899, surrounded by giant sparks and bolts of electricity.

INDUCTION MOTOR
The induction motor has no brushes or commutators. It uses a "rotating magnetic field," made by rapidly feeding carefully timed alternating currents to a series of outer windings – the stator – to produce a magnetic field pattern which rotates. The inner set of windings on the shaft is the rotor. The stator's magnetic field induces a current in the rotor, and this becomes an electromagnet too. As the stator's magnetic field pattern rotates, it "drags" the rotor with it.

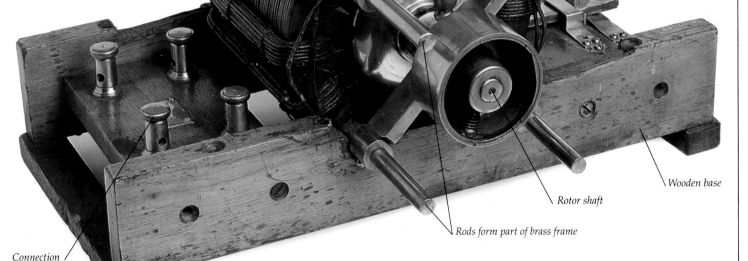

Stator winding

Brass frame

Rotor

Wooden base

Rotor shaft

Rods form part of brass frame

Connection to AC electricity supply

STEPPING UP OR DOWN
The change of voltage in a transformer depends on the number of turns in the two coils, called the primary and secondary windings. If the secondary has more turns than the primary, voltage is increased (stepped up). If it has fewer, voltage is decreased (stepped down). There is no magical gain – as voltage rises, current falls, keeping the overall electrical energy the same.

Step-up transformer

Primary windings to electricity source

Secondary windings connected to high voltage lines

Step-down transformer

Primary windings to electricity source

Secondary windings connected to load – electric motors

TRANSFORMERS IN INDUSTRY
The generators at a modern power station do not produce electricity at sufficiently high voltage for efficient transmission. Their voltage is increased by step-up transformers for efficient long-distance transmission.

Early electricity supplies

THROUGHOUT THE 19TH CENTURY visionaries attempted to put electricity to practical use and even replace steam power, but with no results apart from electroplating (p. 33), some instances of arc lighting (p. 37) and a few small models. In the 1860s efficient generators such as steam turbines were developed, and electricity became available on a larger scale. In the 1880s the rocketing demand for incandescent lamps (pp. 46-47) provided the stimulus for electricity distribution networks. At first, all electricity was generated at the place where it was to be used. This remained the case for a long while until the first central generating stations (pp. 43-44) were built and took over from the small, isolated units.

STEAM ENGINES
James Watt (1736-1819) was a Scottish engineer who made important improvements to the steam engine, thereby helping to stimulate the Industrial Revolution. The unit of power, the watt, is named after him. In electricity, the number of watts is obtained by multiplying volts by amps (p. 25).

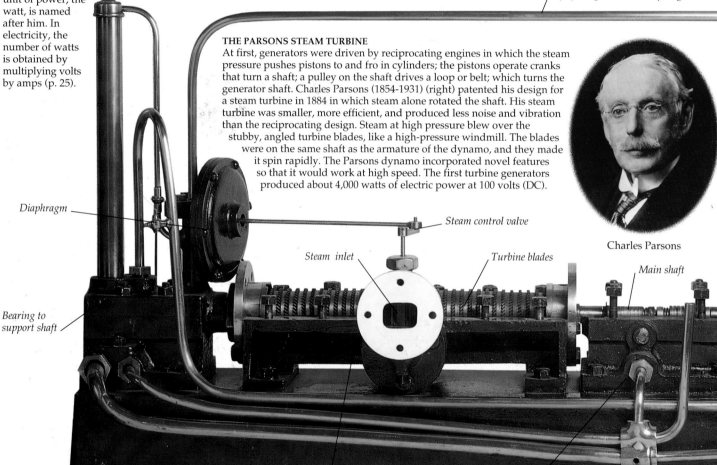

Pipe from governor to diaphragm

THE PARSONS STEAM TURBINE
At first, generators were driven by reciprocating engines in which the steam pressure pushes pistons to and fro in cylinders; the pistons operate cranks that turn a shaft; a pulley on the shaft drives a loop or belt; which turns the generator shaft. Charles Parsons (1854-1931) (right) patented his design for a steam turbine in 1884 in which steam alone rotated the shaft. His steam turbine was smaller, more efficient, and produced less noise and vibration than the reciprocating design. Steam at high pressure blew over the stubby, angled turbine blades, like a high-pressure windmill. The blades were on the same shaft as the armature of the dynamo, and they made it spin rapidly. The Parsons dynamo incorporated novel features so that it would work at high speed. The first turbine generators produced about 4,000 watts of electric power at 100 volts (DC).

Charles Parsons

Diaphragm

Steam control valve

Steam inlet

Turbine blades

Main shaft

Bearing to support shaft

Strong casing (cutaway)

Bearing to support shaft

Electromagnet produces magnetic field

ELECTRICITY FROM STEAM
Inside a steam generator, steam at high pressure is fed through a central inlet into the turbine section, which consists of a spinning shaft inside a strong casing. Half the steam flows in each direction. It rushes through an array of alternate moving turbine blades, which it pushes around, and fixed guide vanes, which guide it in the most efficient way. As steam progresses, its pressure drops but the Parsons turbine distributes this drop in pressure along the blades. The long shaft has a 15-coil armature and commutator mounted near one end. The magnetic field around the armature is produced by the field coils – electromagnets on a horseshoe-shaped iron core.

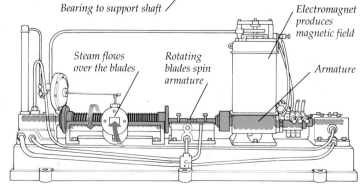

Steam flows over the blades

Rotating blades spin armature

Armature

How the steam turbine works

Electricity to the people

Thomas Edison (1847-1931) began his career as a telegrapher on North American railroads. He turned to inventing, developing things that people might want which he could patent and sell. His first money-maker, in 1870, was an improved stock ticker. This device communicated stock and share prices telegraphically between offices in New York's financial area. There followed a string of inventions and improvements to other people's inventions. Edison was a central figure in the move towards large-scale distribution of electricity to factories, offices, and homes. One of his commercial aims was to break the monopoly of the gas companies, which he considered unfair.

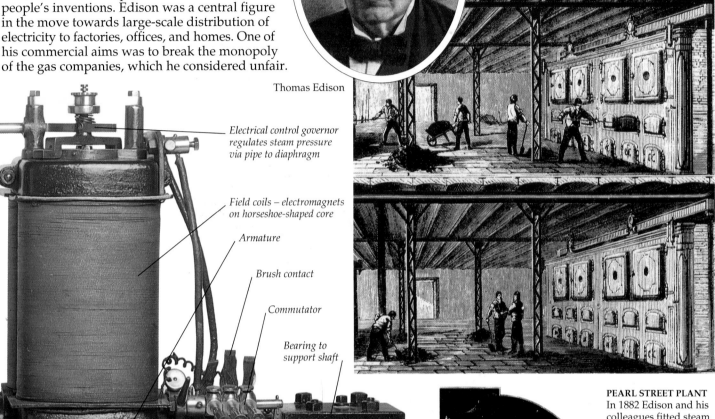

Thomas Edison

Electrical control governor regulates steam pressure via pipe to diaphragm

Field coils – electromagnets on horseshoe-shaped core

Armature

Brush contact

Commutator

Bearing to support shaft

PEARL STREET PLANT
In 1882 Edison and his colleagues fitted steam boilers and improved dynamos into a building on Pearl Street, New York, shown here in this stylized engraving. He had cables installed to distribute the current to the surrounding city district south of Wall Street. Edison manufactured light bulbs and all the other components needed to make it easy for people to equip their homes for electric lighting.

CITY LIGHTS
Before incandescent lamps (p. 47), some city streets were lit by electric arc lamps, as in this New York scene of 1879. But the light was unpleasantly glaring and unreliable (p. 35), so gas was used for most street lighting.

HORSE-DRAWN POWER
The early Parsons turbine generators could be moved by means of a horse-drawn cart to the required site. Their uses were varied. Some were used to provide temporary electric lighting. In 1886 this practical Parson's turbine was used to provide the electric light for ice skating after dark in the northeast of England when a local pond froze.

The power station

AN ELECTRICITY SUPPLY is regarded as a major requirement for a modern society. In the late 19th century electricity-generating power stations were being installed in many of the world's large cities, although it was some decades before the suburbs, and then the rural areas, received supplies. Today, electricity is such a familiar and convenient form of energy that it is simply called "power" – a word with its own scientific meaning (the rate at which energy is used). The majority of modern power stations use turbines to turn generators. These turbines are turned by running water in hydroelectric power stations, or by steam obtained from water boiled by the heat from burning coal, oil, nuclear fuel, or some other source. Less-developed areas may obtain their electricity from local generators.

BIGGER AND BETTER
This model shows the planned Deptford Central Station of the London Electricity Supply Corporation. The engineer who designed it, Sebastian Ferranti (1864-1930), believed it would be more economical to build large power stations outside cities where land was cheap, and to transmit the power at high voltage to substations near the users. Unfortunately, he was ahead of his time. Bureaucracy and technical problems kept the station from being completed. The machines on the right-hand side were installed and supplied power. The enormous ones on the left were never finished.

Flat copper coils

Feedpipes

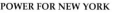

Flywheel

Connecting rod to crank shaft

Check valve

POWER FOR NEW YORK
Great power stations, like this Edison Company station in New York, were often built on riverbanks or coastlines. The coal to fire the boilers could be delivered by barge, and river or sea water supplied the cooling requirements. Electricity changed people's lives, but smoke belched from the chimneys and soot settled over the neighborhood.

DISTRIBUTION NETWORK
A typical power station generator produces alternating current at 25,000 volts. This is stepped up to hundreds of thousands of volts to reduce the energy loss during long-distance transmission. A main substation step-down transformer (pp. 40-41) reduces the voltage for the local area. Smaller substation transformers reduce it further for distribution to offices and homes.

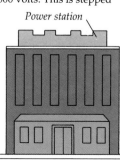

Power station

Looking to the future

The production of electricity is using up nonrenewable sources – coal, oil, and gas. Energy sources that will not run out are now under investigation: solar power, tides, wind, super-heated water gushing from the Earth (geothermal energy), and inflammable gases from "bio-digestors" of rotting plant and animal matter. Nuclear fuel is another option. Each method has its advantages and disadvantages.

Dynamos (behind) are driven by 40 ropes direct from flywheel

Drive pulley or drum

Cylinders

Combined engines and dynamos

Thick concrete floor to support heavy machinery

Feedpipes from boiler

POWER FROM THE SUN
People are now turning to nature's energy source for electric power. This solar power complex in the Mojave Desert in the US generates electricity from rows of computer-controlled mirrors. The mirrors track the sun, and reflect and focus its rays on to tubes containing a special oil. The oil is used to boil water to drive turbines.

HYDROELECTRIC POWER
Glen Canyon dam on the Colorado River in the United States channels running water through great turbines, which turn generators. In countries with sufficient rainfall and plentiful fast rivers, such as New Zealand, hydroelectric stations generate most of the total electricity needs.

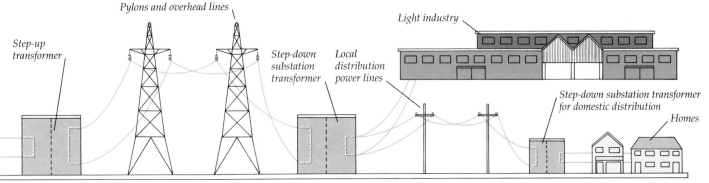

Step-up transformer

Pylons and overhead lines

Step-down substation transformer

Local distribution power lines

Light industry

Step-down substation transformer for domestic distribution

Homes

Electricity in the home

ELECTRICITY WAS FIRST WIRED from power stations into homes, offices, and factories in the 1880s, in big cities such as New York, London, and Paris. Its first major use was for lighting – and it seemed both miraculous and mysterious. Instead of fiddling with gaslights, oil lamps, and candles, users could turn night into day at the flick of a switch. In 1882 Thomas Edison's factories made 100,000 light bulbs; but because of the years it took to lay cables and establish electricity locally, electric light was not easily available until the 1930s. The new power was dangerously invisible. A wire looked the same whether it was carrying a current or not. Early users were warned of the new dangers, though electricity was still safer than the naked flames of candles and gas.

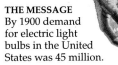

THE MESSAGE
By 1900 demand for electric light bulbs in the United States was 45 million.

Early electric meter

ELECTRICITY METERS
An early meter (left), pioneered by Thomas Edison, used electrolysis (pp. 32-33) to measure the electricity used. The current passed through copper sulfate solution in the jars. The passage of the electric current caused the copper to dissolve on one plate and be deposited on the other. The change in weight of the plates was proportional to the electricity used. The wavy resistance wire and the lamp below prevented the solution from freezing in winter. By the 1930s the meter (below) had a spinning plate moved by induction (pp. 34-35), geared to a series of dials. These showed the amount of electrical energy used, measured in kilowatt-hours (1,000 watts of power for one hour of time).

Jar containing copper sulfate solution

Resistance wire

Copper plates

Lamp to provide heat

1930s meter

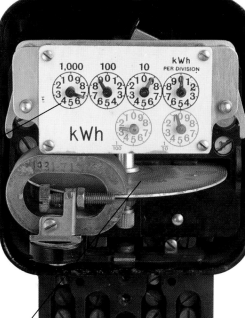

Dials showing kilowatt hours

READING THE METER
Officials from electricity companies were soon visiting consumers on a regular basis to check the electricity meter. The meter reader switched off the supply, removed the copper plate, replaced it with a new one, and took the plate away for weighing. In this way companies figured out how much the consumer should pay.

Spinning plate moved by induction

Electric lamps

Around 1880 light bulbs were developed by Edison, Joseph Swan (1828-1914), and others. These bulbs were called filament or incandescent bulbs, because of the way they worked. Electricity flowed through a thin piece of carbon with high resistance – the filament. This became so hot that it glowed white, or incandesced. If the filament glowed, oxygen in the air would combine with it and quickly burn it away, so air was sucked out of the bulb, to create a partial vacuum around the filament.

Modern light bulb

THE FILAMENT LAMP
The modern light bulb works on the same principle as the Edison-Swan versions, but the materials have changed. The filaments made of tungsten can now withstand very high temperatures. The filament may be 20 in (50 cm) long, but is coiled tightly to take up little space. An inert gas such as argon, to reduce melting of the tungsten, has replaced the vacuum.

Bulb fits into ceramic safety socket

Wires that carry electricity to and from filament

Filament support

Inert gas such as argon

Hard metal (tungsten) filament

FLUORESCENCE
The French scientist Antoine Becquerel (1852-1908) used fluorescent light in his discovery of radioactivity. This type of light works when a substance gives off light after being stimulated by other rays, such as invisible ultraviolet (UV) light. However, practical fluorescent lamps were not in use until the 1950s.

Edison screw-in bulb

Contact metal end

Metal screw fitting

Carbonized bamboo filament

Wires carrying electricity to and from filament

Partial vacuum within bulb

TURNING ON THE LIGHT
This early electric switch has the metal components in a ceramic body. Two metal arms are sprung so they flick quickly from one position to another. When the switch is turned on, the arms slot into U-shaped springy contacts, so completing the electric circuit. This must happen fast, and the contacts must be good, or the electricity could spark across any gaps and cause damage.

Switch turned off

Ceramic body

Circuit broken here

Switch turned on

Sprung metal arms

Electricity flows out

Electricity flows in

Circuit completed

Bayonet bulb

Metal pin forms bayonet fitting

Carbonized cotton filament

Vacuum within bulb

Electrical appliances

As SOON AS ELECTRICITY was available in houses, people began to think up new uses for it. Although the early 20th century saw the invention and design of "labor-saving" appliances to make domestic life easier, the electric iron was the only appliance commonly found in the home, along with electric lights. Most early appliances used electricity's ability to generate heat in appliances like curling irons. When electric motors (pp. 38-39) came into wide use in the 1900s, electricity could be converted into movement. The range of appliances grew to include small heaters, food blenders, and hair dryers. However, the large appliances, such as the vacuum cleaner, were still only found in more affluent homes.

ELECTRIC COOKING
Electric cooking in the 19th century offered freedom from the smoke, burning coals, and hot ashes of the traditional stove. Hot plates could be switched on and off and temperatures adjusted to give the cook greater control.

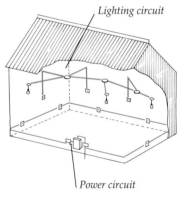

Arc jumps here

Carbon rod

Insulated wire

Wooden handle

Connection for electric cord

Heavy cast-iron base

SMOOTHING IRON
The first electric irons used a high-temperature electric spark for a heat source. This was an an arc, jumping between carbon rods. The rods burned away, so they had to be manually slid together when the electricity was turned off, to maintain the correct gap between them. Like the carbon arc lights, which used the same principle (p. 37), this method of changing electricity into heat was unsafe and unreliable. In 1883 the safety iron was patented in the United States. It replaced the carbon rods with a heating element.

Lighting circuit

Individual fuses   *Indicator light*

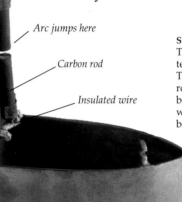

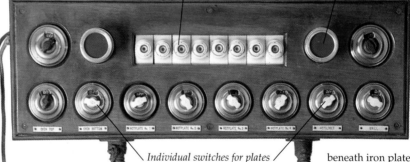

Power circuit

THE HOUSE CIRCUITS
Most modern houses have different electric circuits serving different needs. A main service line leads to a central distribution panel. Separate branch circuits go to wall outlets and lighting fixtures. A heavy-duty circuit serves the kitchen and its appliances. These lines carry 120 volts. Some equipment, such as water pumps, ranges, and dryers, require special circuits with 240 volts.

Wires carrying power to switchboard and range

Individual switches for plates

Wires from switchboard to range encased in corrugated metal

Iron plate

High-resistance wire inside insulated cast-iron case

ELECTRIC RANGE
In this 1900 range, coils of high-resistance wire were placed beneath iron plates. The wire became hot as electricity flowed through it, and the heat passed to the plate. Insulating materials were necessary to prevent electricity from flowing from the wire to the plate.

ELECTRIC KETTLE
Water and electricity are a dangerous combination, since water is an electricity conductor. The first electric kettles had a separate compartment for the element beneath the water. The element heated the container itself but most heat passed into the air. The Swan electric kettle of 1921 was the first with a fully insulated, waterproof heating element in the water.

Copper reflecting dish

Wire safety grill

High-resistance coiled wire

Wooden handle

Heating element

Copper body

Electric motor

Hinge to raise beaters

THE MOTORIZED BLENDER
Mixing, whisking, and beating ingredients by hand is a tiring and time-consuming process. This 1918 food beater and mixer was one of the earliest to be driven by a small electric motor.

THE ELECTRIC PERCOLATOR
One of chief advantages of an electrical appliance is that it can be moved around and used wherever there is a socket to plug it in. This 1912 coffee maker is plugged in to an electric light socket.

Whisks

Supporting base for bowl

Current-carrying wire

ELECTRIC HEATER
The coiled, high-resistance heating element in this 1930s portable heater glows a comforting red as electricity is pushed through it. The copper dish reflects the radiant heat. The types of wire used here show the principle of resistance (pp. 22-23); the high-resistance wire is used as the heating source because it glows hot, while the insulated connecting wires that carry current stay cool and safe to the touch.

Electricity and medicine

THE WATERY SOLUTION OF CHEMICALS in living tissue makes a moderately good electrical conductor. Galvani noted the responses of dissected frog nerves and muscles to discharges from an electrostatic machine (p. 16), and many scientists detected electricity by the shocks it gave them. The body's own "electrical signals" are very small, measured in microvolts, and travel along nerves, sometimes at great speed. They detect, coordinate, and control, especially in the sense organs, brain, and muscles. If increasingly powerful electrical pulses are fed into the body, they cause tingling and then pain; they can send muscles into uncontrolled spasm, burn, render unconscious, and kill. However, doctors have discovered that, used carefully, electricity can diagnose, heal, and cure. Today, electrically heated scalpels slice and then seal small blood vessels to reduce bleeding during operations. Controlled currents passing through tissue can relieve pain. Delicate surgery can be performed with electric laser scalpels.

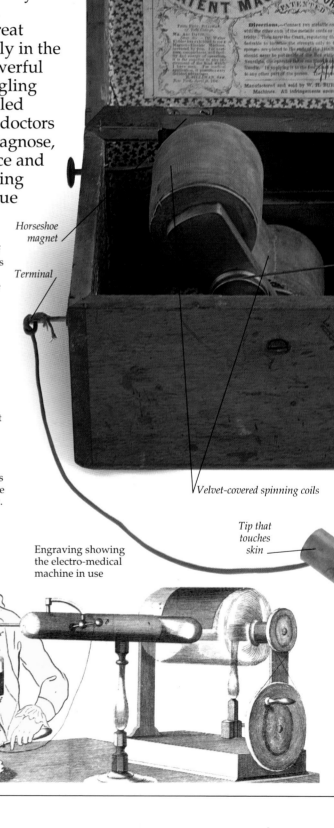

Horseshoe magnet

Terminal

Velvet-covered spinning coils

Tip that touches skin

A CURIOUS TREATMENT
This electro-medical machine of the 1850s had a handle and series of gear wheels which whirled coils of wire next to a horseshoe magnet, to generate electric current in the coils. A simple switch mechanism used this current and the self-inductance of the coils to produce a series of high-voltage pulses which caused the electric shock. The operator held the metal applicators by their insulating wooden handles, and applied the tips to the skin of the patient (below), so the pulses traveled through the tissues to treat rheumatism, headaches, and aching joints. When it was discharged through the patient's tissues, the shock often made the muscles contract uncontrollably.

EXECUTION
Movie villains sometimes meet their end in a dazzling flash of electricity. The reality is the electric chair, and the accidental electrocutions which occur each year in factories, offices, and homes as a result of carelessness.

Engraving showing the electro-medical machine in use

SUBDUING ANIMALS
In the 19th century electricity was sometimes used for controlling vicious horses. A generator was operated by the driver to deliver a shock via a wire in the rein and the metal bit into the horse's mouth.

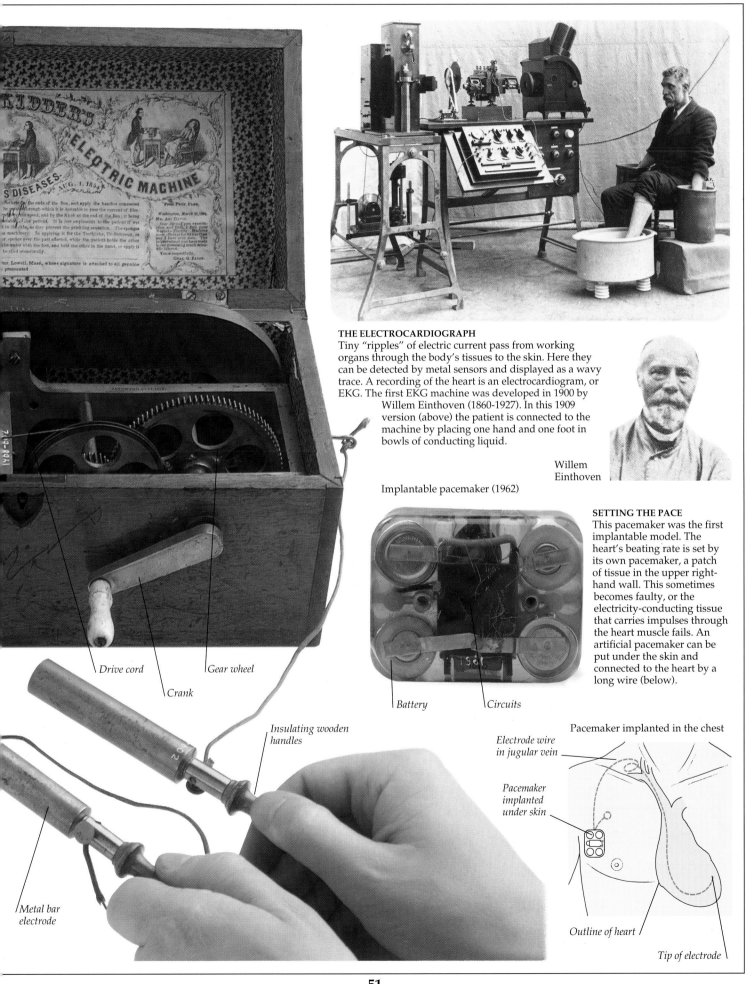

THE ELECTROCARDIOGRAPH

Tiny "ripples" of electric current pass from working organs through the body's tissues to the skin. Here they can be detected by metal sensors and displayed as a wavy trace. A recording of the heart is an electrocardiogram, or EKG. The first EKG machine was developed in 1900 by Willem Einthoven (1860-1927). In this 1909 version (above) the patient is connected to the machine by placing one hand and one foot in bowls of conducting liquid.

Willem Einthoven

Implantable pacemaker (1962)

SETTING THE PACE

This pacemaker was the first implantable model. The heart's beating rate is set by its own pacemaker, a patch of tissue in the upper right-hand wall. This sometimes becomes faulty, or the electricity-conducting tissue that carries impulses through the heart muscle fails. An artificial pacemaker can be put under the skin and connected to the heart by a long wire (below).

Drive cord

Gear wheel

Crank

Battery

Circuits

Insulating wooden handles

Metal bar electrode

Pacemaker implanted in the chest

Electrode wire in jugular vein

Pacemaker implanted under skin

Outline of heart

Tip of electrode

Heat, pressure, and light

ELECTRICITY CAN BE PRODUCED DIRECTLY from heat, pressure, and light. When heat is applied to one of the junctions of two conductors, so that the two junctions are at different temperatures, an electrical potential is generated; this is thermoelectricity. In the piezoelectric effect, an electrical potential is generated between opposite faces of crystals made from substances such as quartz, when the crystals are compressed – in simple terms, when the crystal is squeezed or stretched. In the photoelectric effect, light rays cause certain substances to give up electrons (p. 20) and produce an electric charge or current. In the photovoltaic effect, light produces an electric potential between layers of different substances, so current flows in a circuit without need for an electricity source.

THOMAS SEEBECK
In the 1820s German scientist Thomas Seebeck (1770-1831) studied the effects of heat on conductors. This "Seebeck effect" is now known as thermoelectricity. There have been many attempts at using thermoelectricity. Its main use today is for thermometers (below right).

THE SEEBECK EFFECT
In a circuit of two metal strips joined at their ends, Seebeck heated the junction at one end and saw a compass needle between the strips swing. Seebeck thought the heat was creating magnetism. In fact, an electric current was generated when the junctions were at different temperatures, and the magnetic field it produced deflected the compass needle. The Seebeck effect is effective only with certain conductors. Seebeck obtained his best results with the dissimilar conductors bismuth and antimony. The same effect can be produced using iron and copper. The "opposite" is the Peltier effect, named after French scientist Jean Peltier (1785-1845). A temperature difference is produced between the junctions when an electric current from a battery flows around the circuit.

Reconstruction of Seebeck's experiment

Copper strip

Junction at room temperature

Deflected compass needle

Current flows around circuit

Junction being heated

Metal base

Iron strip

Bunsen burner

MAKING A THERMOCOUPLE
One application of the Seebeck effect is the type of thermometer based on a thermocouple (opposite). Here the probe of the thermocouple is being made from a wire of platinum joined by welding to a wire of platinum-rhodium alloy.

PIEZOELECTRIC EFFECT

In the crystal pickup of a record player, mechanical vibrations produced in the stylus by the grooves in the record are transmitted to the piezoelectric crystal by a plastic stirrup. This produces a varying electrical signal to match the intended sound. The vibrations are directed in two directions, at right angles, to make the two channels necessary for stereophonic sound. The signal is amplified and then fed to loudspeakers (p. 60). Other applications of piezoelectricity include the pager and hand-squeezed spark-makers used for both cigarette and gas oven lighters.

PIERRE CURIE

The piezoelectric effect was first studied in about 1880 by Pierre Curie (1859-1906) and his brother Jacques (1859-1941). It is named from the Greek word *piezein* meaning to press. The Curies used Rochelle salt and quartz, natural crystals with piezoelectric qualities. Nowadays synthetic crystals are used.

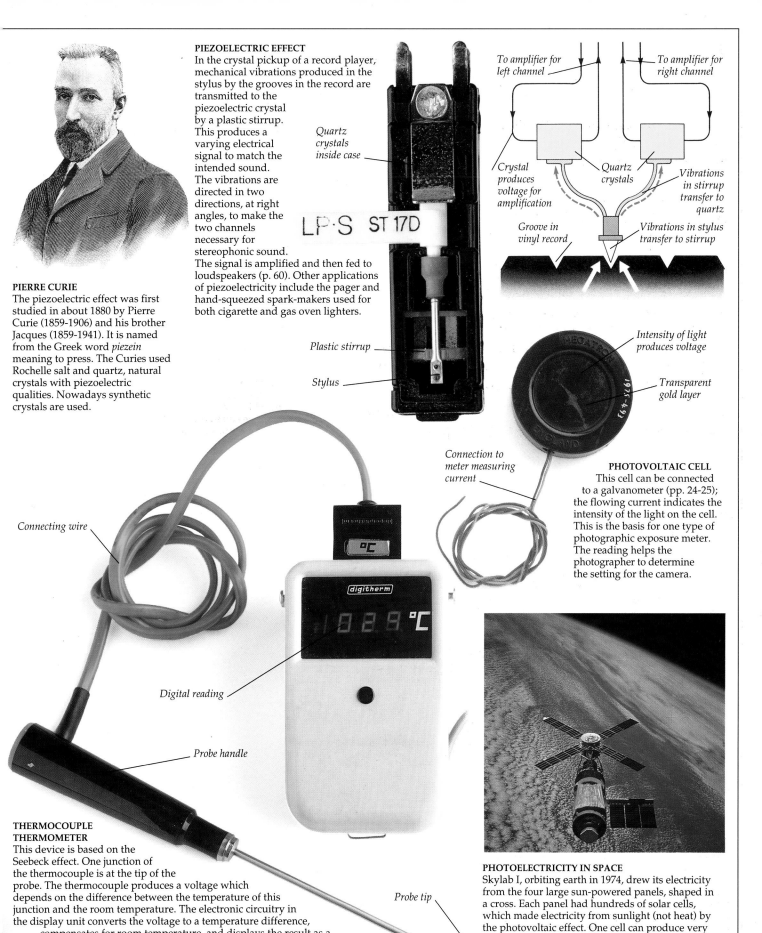

Quartz crystals inside case

LP·S ST 17D

Plastic stirrup

Stylus

To amplifier for left channel

To amplifier for right channel

Crystal produces voltage for amplification

Quartz crystals

Vibrations in stirrup transfer to quartz

Groove in vinyl record

Vibrations in stylus transfer to stirrup

Intensity of light produces voltage

Transparent gold layer

Connection to meter measuring current

PHOTOVOLTAIC CELL

This cell can be connected to a galvanometer (pp. 24-25); the flowing current indicates the intensity of the light on the cell. This is the basis for one type of photographic exposure meter. The reading helps the photographer to determine the setting for the camera.

Connecting wire

Digital reading

digitherm

Probe handle

THERMOCOUPLE THERMOMETER

This device is based on the Seebeck effect. One junction of the thermocouple is at the tip of the probe. The thermocouple produces a voltage which depends on the difference between the temperature of this junction and the room temperature. The electronic circuitry in the display unit converts the voltage to a temperature difference, compensates for room temperature, and displays the result as a digital reading. This thermometer uses a nickel-chromium and nickel aluminum thermocouple and measures temperatures from –58° F to 1830° F. Other types have a wider range. These thermometers have replaced the mercury type for many uses.

Probe tip

PHOTOELECTRICITY IN SPACE

Skylab I, orbiting earth in 1974, drew its electricity from the four large sun-powered panels, shaped in a cross. Each panel had hundreds of solar cells, which made electricity from sunlight (not heat) by the photovoltaic effect. One cell can produce very little electrical energy, so the cells were connected together to make useful amounts of electricity.

Investigating cathode rays

IT WAS KNOWN IN THE 18TH CENTURY that a gas at low pressure inside a tube could be made to glow by passing a discharge from an electrostatic machine through it. In the latter part of the 19th century new electrical apparatus enabled scientists to study the effect more thoroughly. For instance, improved vacuum pumps allowed the air pressure inside the tubes to be reduced. At very low pressures the glow disappeared and was replaced by invisible rays which came from the cathode or negative terminal. This made the glass of the tube containing the gas glow green where the rays struck it. In 1883, Edison had noticed that particles seemed to be given off from the negative ends of the filaments of his electric light bulbs. The British scientist William Crookes was one of the discoverers of these "cathode rays." Later, J. J. Thomson extended this work with insights that led to the new science of atomic physics and to the discovery of the structure of the atom by Ernest Rutherford (1871-1937) and his team. The cathode ray tube became the basis for the television set (pp. 62-63) and radar equipment.

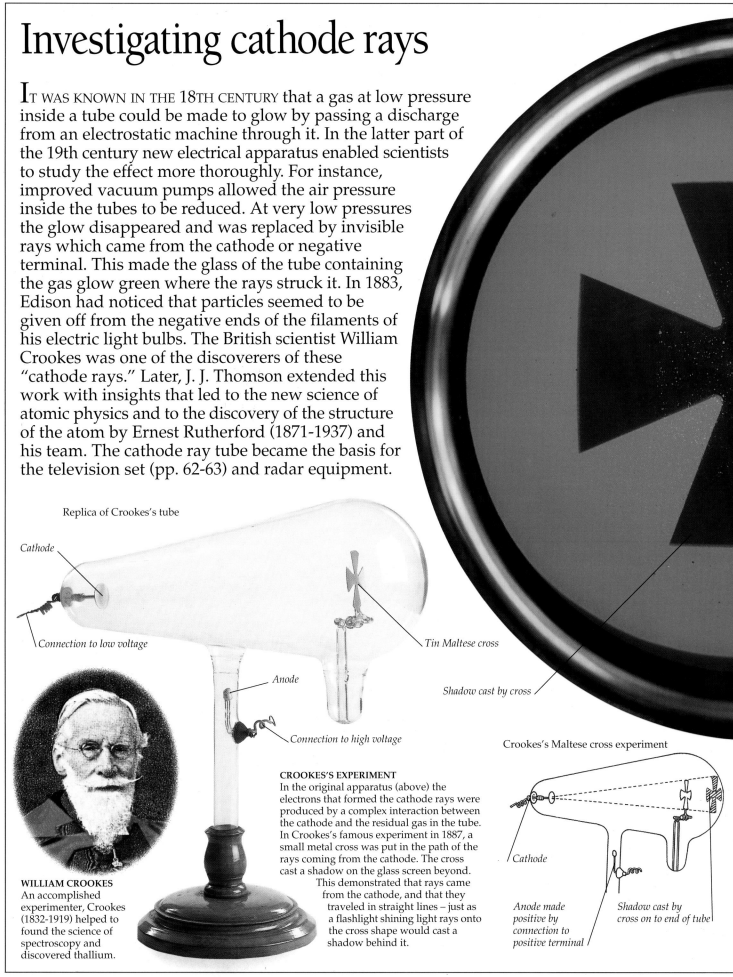

Replica of Crookes's tube

Cathode

Connection to low voltage

Tin Maltese cross

Shadow cast by cross

Anode

Connection to high voltage

WILLIAM CROOKES
An accomplished experimenter, Crookes (1832-1919) helped to found the science of spectroscopy and discovered thallium.

CROOKES'S EXPERIMENT
In the original apparatus (above) the electrons that formed the cathode rays were produced by a complex interaction between the cathode and the residual gas in the tube. In Crookes's famous experiment in 1887, a small metal cross was put in the path of the rays coming from the cathode. The cross cast a shadow on the glass screen beyond. This demonstrated that rays came from the cathode, and that they traveled in straight lines – just as a flashlight shining light rays onto the cross shape would cast a shadow behind it.

Crookes's Maltese cross experiment

Cathode

Anode made positive by connection to positive terminal

Shadow cast by cross on to end of tube

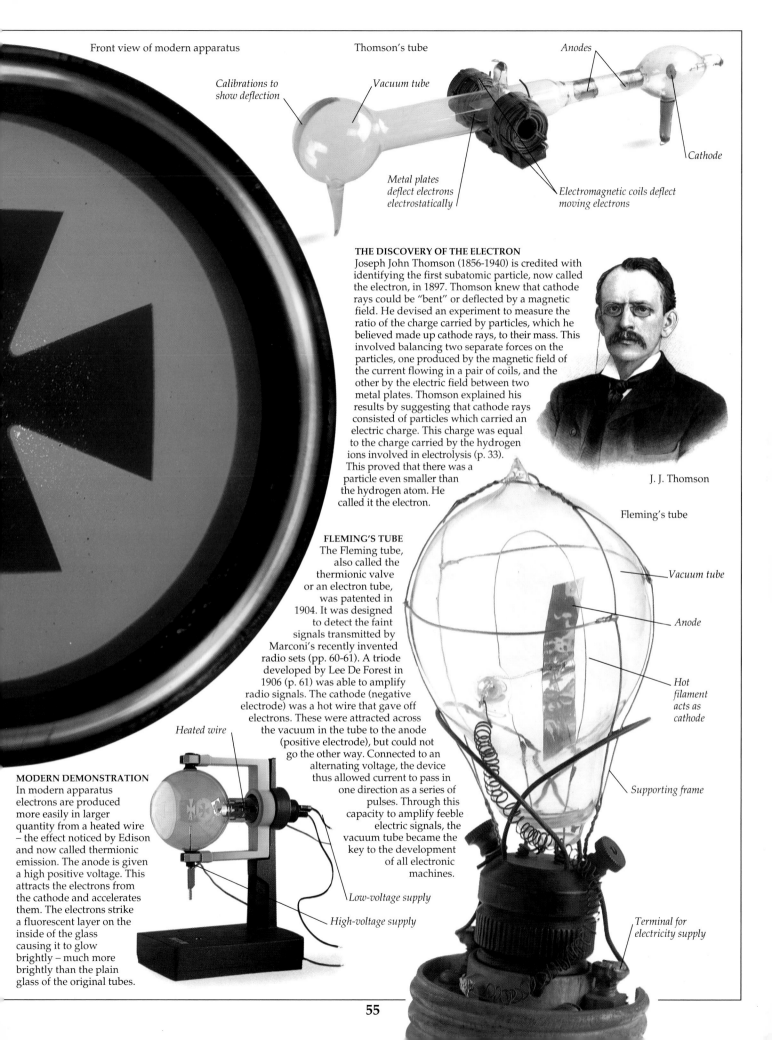

Front view of modern apparatus

Thomson's tube

Calibrations to show deflection

Vacuum tube

Anodes

Metal plates deflect electrons electrostatically

Electromagnetic coils deflect moving electrons

Cathode

THE DISCOVERY OF THE ELECTRON

Joseph John Thomson (1856-1940) is credited with identifying the first subatomic particle, now called the electron, in 1897. Thomson knew that cathode rays could be "bent" or deflected by a magnetic field. He devised an experiment to measure the ratio of the charge carried by particles, which he believed made up cathode rays, to their mass. This involved balancing two separate forces on the particles, one produced by the magnetic field of the current flowing in a pair of coils, and the other by the electric field between two metal plates. Thomson explained his results by suggesting that cathode rays consisted of particles which carried an electric charge. This charge was equal to the charge carried by the hydrogen ions involved in electrolysis (p. 33). This proved that there was a particle even smaller than the hydrogen atom. He called it the electron.

J. J. Thomson

Fleming's tube

FLEMING'S TUBE

The Fleming tube, also called the thermionic valve or an electron tube, was patented in 1904. It was designed to detect the faint signals transmitted by Marconi's recently invented radio sets (pp. 60-61). A triode developed by Lee De Forest in 1906 (p. 61) was able to amplify radio signals. The cathode (negative electrode) was a hot wire that gave off electrons. These were attracted across the vacuum in the tube to the anode (positive electrode), but could not go the other way. Connected to an alternating voltage, the device thus allowed current to pass in one direction as a series of pulses. Through this capacity to amplify feeble electric signals, the vacuum tube became the key to the development of all electronic machines.

Vacuum tube

Anode

Hot filament acts as cathode

Supporting frame

Heated wire

MODERN DEMONSTRATION

In modern apparatus electrons are produced more easily in larger quantity from a heated wire – the effect noticed by Edison and now called thermionic emission. The anode is given a high positive voltage. This attracts the electrons from the cathode and accelerates them. The electrons strike a fluorescent layer on the inside of the glass causing it to glow brightly – much more brightly than the plain glass of the original tubes.

Low-voltage supply

High-voltage supply

Terminal for electricity supply

Communicating with electricity

TELEGRAPHY – WRITING AT A DISTANCE – was an early and successful use of electricity, coinciding with the spread of railroads in Europe and North America. Telegraph wires or lines were laid alongside the railroad tracks, where they were easy to check and maintain. The discoveries about the connection between electricity and magnetism (pp. 26-27) had led several people to develop the idea of sending pulses of current in coded form from one place to another. A simple on-off switch completed the circuit and allowed current to flow from the sender. This was read in a code of dots and dashes. Although it took several years for the electric telegraph to be accepted, the era of fast communication had begun.

THE WHEATSTONE TELEGRAPH
English experimenters William Cooke (1806-1879) and Charles Wheatstone (1802-1875) demonstrated a telegraph in England in 1837. At first people were suspicious of the electric wires passing over their land. Wheatstone used the deflections of a needle to spell out letters, in a code of swings to the left and right. The needle was moved by the magnetic field produced by the current through coils of wire inside the case. Twisting the handle one way connected the positive terminal of the battery into the circuit and the negative one to ground, sending a current in one direction, making the needle of the receiving instrument deflect one way. Turning the handle the other way connected the negative terminal of the battery, making the needle deflect the other way.

Single-needle instrument with back removed (1846)

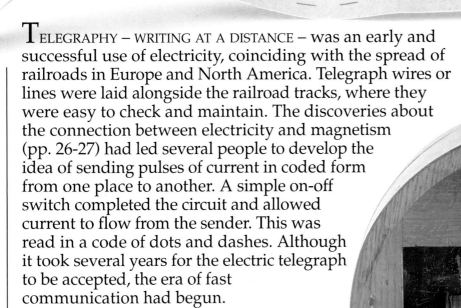

Coils of wire

Operating handle

Magnetic needle that acts as pointer

THE TRAIN IS COMING
Along a copper wire, the signal travels at nearly the speed of light, making telegraphy an almost instant method of sending messages. It was used extensively for signalling on the railroads. This operator listens to the clicking code through earphones; the batteries are on the floor under his desk.

Wires to telegraph

Trough battery   *Wire to battery*

A .–	J .–––	S ...	2 ..–––
B –...	K –.–	T –	3 ...––
C –.–.	L .–..	U ..–	4 ....–
D –..	M ––	V ...–	5 .....
E .	N –.	W .––	6 –....
F ..–.	O –––	X –..–	7 ––...
G ––.	P .––.	Y –.––	8 –––..
H	Q ––.–	Z ––..	9 ––––.
I ..	R .–.	1 .––––	0 –––––

Modern Morse code alphabet

The "father of the telegraph"

American Samuel Morse (1791-1872) began his career as a professional portrait painter. The idea for a telegraph came to him in 1832 when he saw an electromagnet on a ship while returning from a European tour. He gave up painting and began to investigate electromagnetism. By 1837 he had devised electromagnetic transmitters and receivers, and the first version of the code of dots and dashes that took his name and eventually became used worldwide. Morse's first permanent telegraph line, spanning 37 miles (60 km) between Baltimore and Washington, opened on May 24, 1844 with Morse's message, "What hath God wrought!"

Samuel Morse

DOTS AND DASHES

Morse's early designs for a receiver used an electromagnet and a stylus that pressed a groove into a moving paper strip. Later, there was an inking device (p. 31) for writing dots and dashes. Trained telegraphers listened to the receiver's clicks or a sounding device, decoded the message from these, and wrote it down.

Morse tapes received on the *Great Eastern* when it was laying the 1865 Atlantic cable

Battery

Relay

Transformer

Electromagnets

Coherer

WIRELESS RECEIVER

An advance on telegraphy, based on electromagnetic waves, was the wireless telegraph. This was made possible by the early experiments of Heinrich Hertz (pp. 60-61) and the practical efforts of Guglielmo Marconi (p. 61). This wireless receiver was the same as the telegraphy receivers in that it used Morse code, but instead of a connection to a land line, it had a radio receiver which used a coherer. The coherer contains filings that increase in conductivity when they are subjected to radio waves. The link between the radio signal and the sounder circuit is made by the relay, which uses electromagnets (pp. 30-31). It passes the on-off pattern of the received current to a new, more powerful circuit.

Wire to battery

LONG-DISTANCE BUSINESS

Telegraph lines spread across the land. Their real importance was appreciated by businesses where investments and information could be received quickly. These wealthy businessmen no doubt await the 19th-century version of the latest stock market figures.

Talking with electricity

SOUND WAVES DO NOT TRAVEL VERY FAR or very fast in air. As the telegraph system became established for sending messages in coded form, several inventors pursued the idea of using the complex pattern of sound waves from the voice to produce a corresponding pattern of electric signals. These could be sent along a wire much farther, and faster, than sound waves in the air. At the other end, the electric signals would be converted back to sound, to recreate the original speech. A receiver and transmitter at each end allowed both callers to speak and listen. Bell, Edison, and others succeeded in making such devices, which became known as telephones. Telegraphs retained their role of sending relatively simple messages.

ALEXANDER GRAHAM BELL
Born in Scotland, Alexander Graham Bell (1847-1922) emigrated to the United States in 1871, where he became an outstanding figure in the education of the deaf. He found that different voice tones could vary the electrical signals flowing in a wire, by the process of electro-magnetic induction. He also realized that a varying signal could vibrate a flat sheet, or diaphragm, and produce sound waves. The principle of the telephone was born.

INSIDE A BELL TELEPHONE
The "candlestick" design (1878) came about due to the mistaken belief that a longer permanent magnet would be stronger than a short one. The same design (below) could function as a mouthpiece (the sound detector or microphone) or earpiece (the sound producer). Sound waves vibrate an iron diaphragm varying the magnetic field of the bar magnet, which produces varying currents in the coil by electromagnetic induction. The principle of the earpiece design changed little from Bell's time until the 1970s, though improved materials allowed the permanent magnet inside the earpiece to be smaller. It is Edison's carbon microphone that is used for the typical mouthpiece.

Terminals

Bell's first telephone

THE FIRST CALL
With advice from Joseph Henry (p. 35), Bell and his assistant, Thomas Watson, constructed early versions of the telephone. Sound waves from the speaker's voice funnelled into a chamber, where they vibrated a flat sheet of thin iron, the diaphragm. This disturbed the magnetic field of a permanent magnet, around which was wrapped a current-carrying coil connected to an external battery. The magnetic field induced varying electrical signals in the circuit, which was completed through a grounded connection.

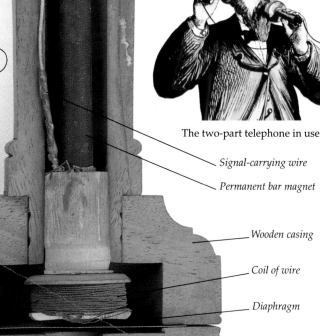

The two-part telephone in use

Signal-carrying wire

Permanent bar magnet

Wooden casing

Coil of wire

Diaphragm

Funnel for sound waves

OPERATOR SERVICE
Only a few years after the first telephones were demonstrated, exchanges were being set up in major cities. The caller's plug had to be inserted into the correct socket. This is Croydon Exchange, near London, which opened in 1884.

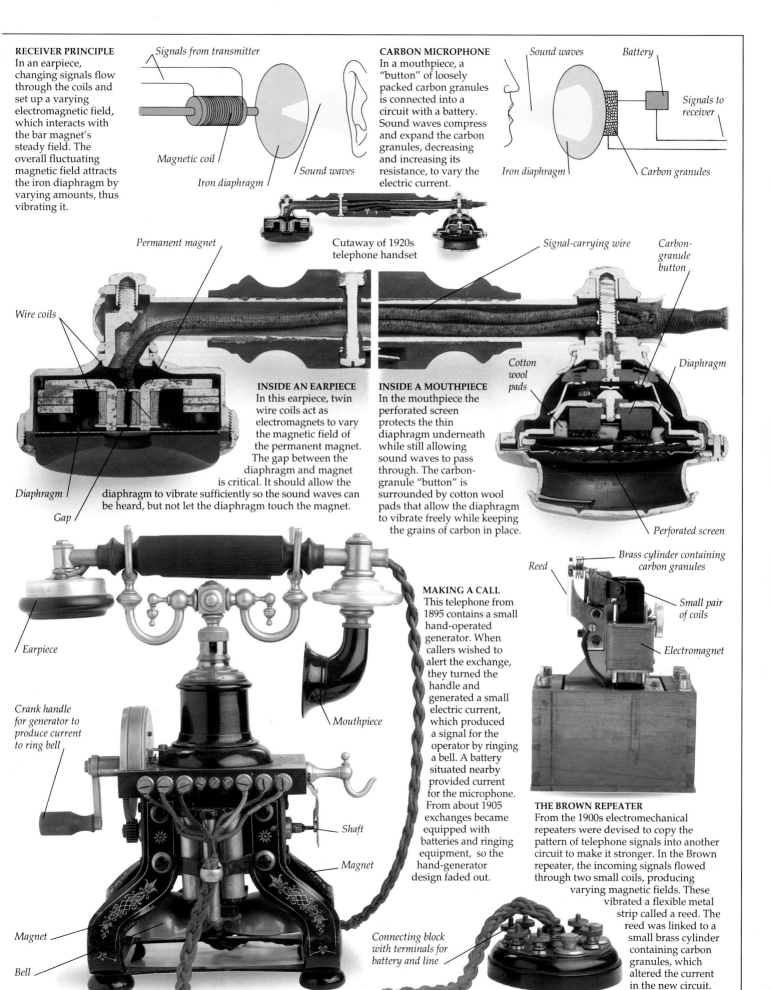

RECEIVER PRINCIPLE
In an earpiece, changing signals flow through the coils and set up a varying electromagnetic field, which interacts with the bar magnet's steady field. The overall fluctuating magnetic field attracts the iron diaphragm by varying amounts, thus vibrating it.

Signals from transmitter

Magnetic coil

Iron diaphragm

Sound waves

CARBON MICROPHONE
In a mouthpiece, a "button" of loosely packed carbon granules is connected into a circuit with a battery. Sound waves compress and expand the carbon granules, decreasing and increasing its resistance, to vary the electric current.

Sound waves

Battery

Signals to receiver

Iron diaphragm

Carbon granules

Cutaway of 1920s telephone handset

Permanent magnet

Signal-carrying wire

Carbon-granule button

Wire coils

Cotton wool pads

Diaphragm

Diaphragm

Gap

INSIDE AN EARPIECE
In this earpiece, twin wire coils act as electromagnets to vary the magnetic field of the permanent magnet. The gap between the diaphragm and magnet is critical. It should allow the diaphragm to vibrate sufficiently so the sound waves can be heard, but not let the diaphragm touch the magnet.

INSIDE A MOUTHPIECE
In the mouthpiece the perforated screen protects the thin diaphragm underneath while still allowing sound waves to pass through. The carbon-granule "button" is surrounded by cotton wool pads that allow the diaphragm to vibrate freely while keeping the grains of carbon in place.

Perforated screen

Earpiece

Reed

Brass cylinder containing carbon granules

Small pair of coils

Electromagnet

Crank handle for generator to produce current to ring bell

MAKING A CALL
This telephone from 1895 contains a small hand-operated generator. When callers wished to alert the exchange, they turned the handle and generated a small electric current, which produced a signal for the operator by ringing a bell. A battery situated nearby provided current for the microphone. From about 1905 exchanges became equipped with batteries and ringing equipment, so the hand-generator design faded out.

Mouthpiece

Shaft

Magnet

THE BROWN REPEATER
From the 1900s electromechanical repeaters were devised to copy the pattern of telephone signals into another circuit to make it stronger. In the Brown repeater, the incoming signals flowed through two small coils, producing varying magnetic fields. These vibrated a flexible metal strip called a reed. The reed was linked to a small brass cylinder containing carbon granules, which altered the current in the new circuit.

Magnet

Bell

Connecting block with terminals for battery and line

Communicating without wires

JAMES CLERK MAXWELL (1831-1879) developed the findings of Faraday and other scientists, using concepts such as magnetic lines of force, and reduced the phenomena of electricity and magnetism to a group of four mathematical equations. One prediction from these equations was that an oscillating electric charge would send out "waves" of electromagnetic energy from its source. A series of experiments by Heinrich Hertz (1857-1894) in the 1880s demonstrated that these waves existed, and that they could be detected at a distance. Further work by Guglielmo Marconi (1874-1937) in the 1890s resulted in radio telegraphy, the sending of messages without wires.

JAMES CLERK MAXWELL
Scottish-born Maxwell predicted the existence of radio waves before they were demonstrated by Hertz (below). He showed that an oscillating electric charge would produce a varying electromagnetic field, which would transmit at a speed that turned out to be equal to the speed of light. From this, he suggested that light rays were electro-magnetic waves.

HERTZ'S EXPERIMENTS
Hertz demonstrated the existence of radio waves in the late 1880s. He used a device called an induction coil to produce a high voltage. One of his early transmitters consisted of two tiny coils with a spark gap. The rapidly oscillating current in the sparks between the ends of the coils produced radio waves. To detect the waves, Hertz used a receiver consisting of two rods with a spark gap as the receiving antenna. A spark jumped the gap where the waves were picked up. Hertz showed that these signals had all the properties of electromagnetic waves. They could be focused by curved reflectors, and a grid of parallel wires placed in their path showed they were polarized.

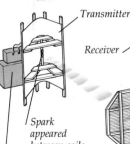

Heinrich Hertz

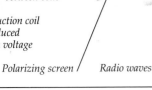

Transmitter

Receiver

Spark appeared between coils

Induction coil produced high voltage

Polarizing screen

Radio waves

Plane reflector

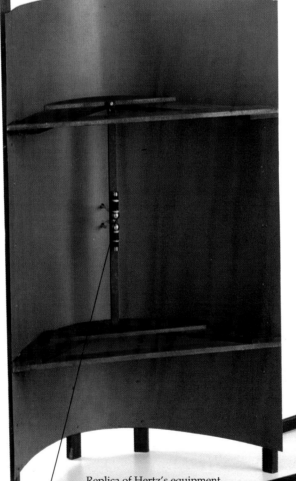

Replica of Hertz's equipment

Two coils and spark gap

Reflector

Polarizing screen

GUGLIELMO MARCONI

Marconi began radio experiments on his family's estate near Bologna, Italy. He devised arrangements of long wires and metal plates to send and receive the waves. These were the first antennas. His first radio message across the Atlantic, from Cornwall in England to Newfoundland on December 12, 1901, was Morse code for the letter S. It established the viability of radio for long-distance communication. Here, Marconi (left) shows some of his equipment to visitors in 1920.

Cone made from cardboard

Permanent magnet and coil in casing

Loudspeaker

How a loudspeaker works

Cylindrical magnet produces strong magnetic field

Varying currents passing through coil of wire vibrate cone, causing sound waves

ELECTRICITY INTO SOUND ENERGY

Headphones were used to listen to the first radio sets. Loudspeakers were then developed so that several people could listen at once. Loudspeakers needed radios with better detectors and amplifiers. This became possible when the triode valve was devised in 1906 by Lee De Forest (1873-1961) since the valve could amplify weak signals. The loudspeaker has a strong permanent magnet shaped like a cylinder. A coil of wire fits between the poles of the magnet within the strong magnetic field, and is attached to the cardboard cone. The varying electric currents pass through the coil making it move to and fro. As the coil moves, the cone moves, making sound waves that correspond to the varying electric currents.

PAPAL AUDIENCE

From the 1920s radio broadcasts became used for spreading news and for entertainment. Pope Pius XI (reigned 1922-1939) listens to a musical broadcast on his headphones.

RADIO RECEIVER

This state-of-the-art radio dates from 1925 and shows that large multi-wound antennas were required to gather the energy represented by radio waves. Each stage of circuitry in the receiver had to be individually tuned to the radio station.

Antenna

The antenna frame can be rotated to produce the strongest signals

Receiver in curved reflector

Rods pick up radio waves

Volume control

Tuning knob

Headphone socket

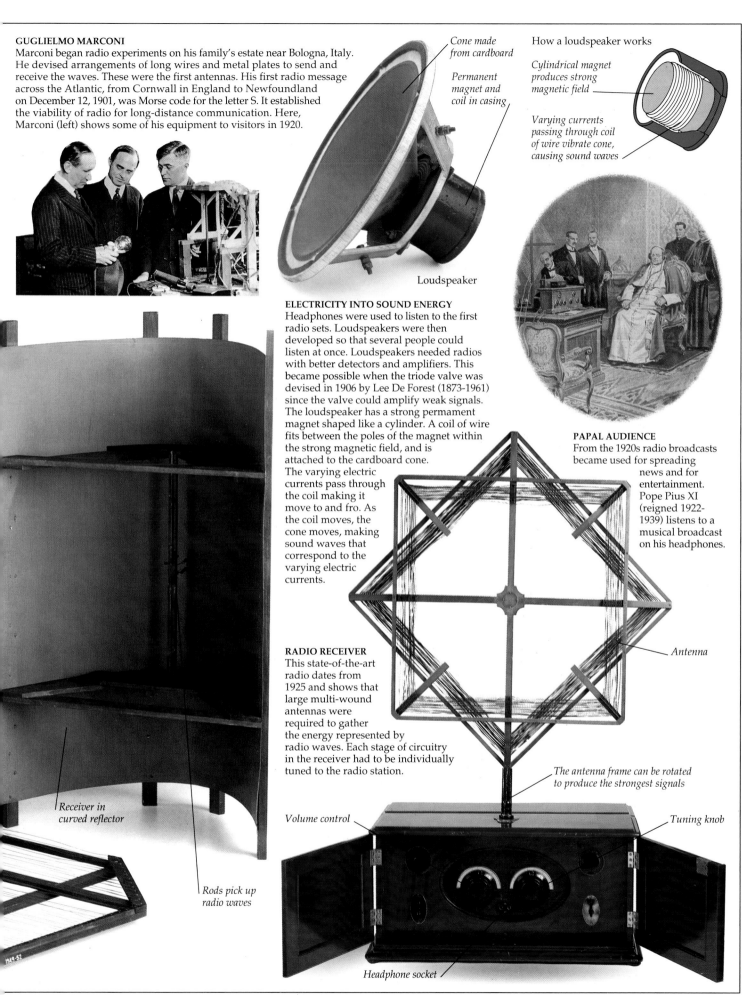

61

Television and the future

In the area of communication, there was another great aim for researchers and inventors. This was the wireless transmission not only of sounds, but of images too – television. Several systems were tried for turning patterns of light into electric signals in the camera, transmitting and receiving the signals as radio waves, and displaying the received signals as a moving image for the viewer. A version of the vacuum tube, the cathode ray tube (CRT), became established as the image-displaying unit. This device was yet another stage in the two centuries of research, manipulation, and utilization of electricity – a form of energy that has become master and servant, indispensible and central to modern scientific thought.

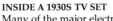

INSIDE A 1930S TV SET
Many of the major electrical components described in this book, or their descendants, are contained in this electric machine. The transformer and associated components produce high voltages which drive electrons in beams from the cathode, through the anodes and towards the screen (pp. 54-55). The screen has a special coating so that it fluoresces white when the electrons hit it. Variable capacitors (p. 13) and other items tune the set to receive signals from different transmitters.

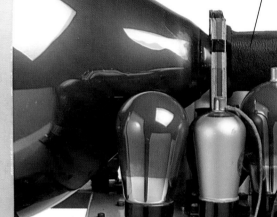

Deflection coils

Screen

Cathode ray tube

Control knobs

Loudspeaker

Condenser

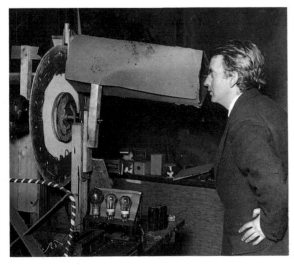

BAIRD'S MECHANICAL SCANNER
The television transmission equipment devised in the 1920s by John Logie Baird (1888-1946) was built using various bits of scrap metal and electrical components. Central was the Nipkow disc, a fast-spinning disc with holes arranged in a spiral. As the disc turned, each hole traced a curved line and exposed part of the scene behind. A photoelectric cell (p. 53) transformed the light intensity of each part of the line into electric signals, and sent them to the receiver. Baird's 1926 version produced a 30-line image which was renewed 10 times each second. This electromechanical system was replaced by a purely electrical one (opposite).

HOW THE TELEVISION WORKS

In the black-and-white television, electrons are produced at the cathode and are accelerated towards the positive electrode, the anode. They pass through holes in the anode, and are focused by the magnetic field produced by focusing coils to produce a spot on the screen. These pairs of focusing coils, one pair arranged vertically and the other horizontally, create fields which deflect the electron beam. The fields are varied so the beam sweeps across the screen, jumps back, sweeps an adjacent line, and so on to cover the screen. At the same time, the intensity of the beam is varied by a signal applied to another electrode near the cathode. When the beam is stronger, it makes the screen glow more brightly at the spot it hits. Each second, 30 complete pictures or frames are produced. The human eye cannot follow the rapid movement of the electron beam and perceives a smoothly changing picture.

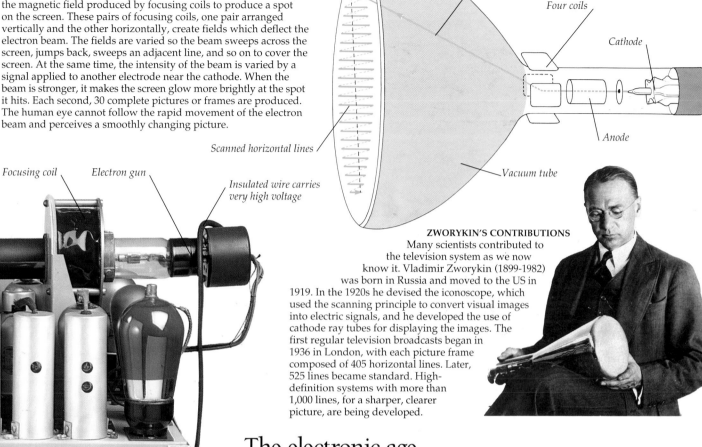

Electron beam

Four coils

Cathode

Anode

Vacuum tube

Scanned horizontal lines

Focusing coil *Electron gun*

Insulated wire carries very high voltage

Resistor *Transformer*

ZWORYKIN'S CONTRIBUTIONS
Many scientists contributed to the television system as we now know it. Vladimir Zworykin (1899-1982) was born in Russia and moved to the US in 1919. In the 1920s he devised the iconoscope, which used the scanning principle to convert visual images into electric signals, and he developed the use of cathode ray tubes for displaying the images. The first regular television broadcasts began in 1936 in London, with each picture frame composed of 405 horizontal lines. Later, 525 lines became standard. High-definition systems with more than 1,000 lines, for a sharper, clearer picture, are being developed.

The electronic age

An understanding of the behavior and nature of electrons led to electronic components such as the valve (p. 55). This was an enormous breakthrough in electrical technology but it relied on heat, wasting lots of energy, and it was relatively fragile. During the late 1940s the first transistors were developed. Transistors were used for similar purposes as valves, and soon replaced them for many applications. In the 1960s techniques were introduced to etch whole networks of components on to a thin wafer or "chip" of silicon, to make integrated circuits.

DIODE VALVE
The experimental Fleming diodes of 1904 were the forerunners of other types of vacuum tube, the triode and pentode, with more electrodes. Valves could use small electric signals and control the current through the valve, and could thus amplify the radio signal (pp. 60-61).

TRANSISTOR
Like valves, transistors could amplify electric signals and manipulate currents and voltages in various ways. But they were smaller, more efficient, more robust, and eventually cheaper than valves.

SILICON CHIP
A tiny wafer of silicon has circuits containing hundreds of transistors and other electrical components.

Wafer of silicon

Index

Acknowledgments

Dorling Kindersley would like to thank:
Fred Archer, Brian Bowers, Roger Bridgman, Janet Carding, Eryl Davies, Robert Excell, Graeme Fyffe, Derek Hudson, Dr Ghislaine Lawrence, Barry Marshall and the staff of the Museum Workshop, Douglas Millard, Victoria Smith, Peter Stephens, Peter Tomlinson, Kenneth Waterman, Anthony Wilson, and David Woodcock for advice and help with the provision of objects for photography at the Science Museum; Dave Mancini at Nortech and Peter Griffiths for the model making; Deborah Rhodes for page makeup, Peter Cooling for computer artwork; Robert Hulse for standing in for Michael Faraday and Humphry Davy; Karl Adamson and Tim Ridley for assistance with the photography; Jack Challoner for help in the initial stages of the book; Susannah Steel for proofreading.

Picture research Deborah Pownall and Catherine O'Rourke
Illustrations Kuo Kang Chen
Index Jane Parker

Picture credits

t=top b=bottom c=center l=left r=right

Ashmolean Museum, Oxford 8tr.
Bildarchiv Preussischer Kulturbesitz 40br.
Bridgeman Art Library 8bc; /Philips 32tl.
Bt Museum, London 8c; 58bl.
Mary Evans Picture Library 7cl; 15bl; 19tc; 19c; 27tc; 30tr; 41tr; 43tr; 44bl; 46tl; 46bl; 49cr; 61tr.
Hulton Picture Co. /Bettmann Archive 14tc; 16bc; 28tl; /Bettmann Archive 52tr; 56cl; 57br; 58cl; 58tl; 61tl; 62cl; /Bettmann Archive 63cr.
Patsy Kelly 12-13.
Kobal Collection 12bl; 50cl.
Mansell Collection 6cl; 24cl; 47cr.
MIT Museum, Cambridge, Massacchusetts 12c.
National Portrait Gallery, London 10tr.
National Portrait Gallery, Washington 29tr.
NEI Parsons 42cr; 43bl.
Ann Ronan Picture Library 11tc; 19cb; 37cl; 42tl; 51tr; 53tl.

Rover Group 33br.
Royal Institution, London 35tr; 35cr.
Science Museum Photo Library cover front tl; back tl; 9tr; 18tl; 23tl; 27clb; 29br; 34tl; 41tc; 48tr; 55cr; 57tr; 60clb.
Science Photo Library /Gordon Garrado cover front c; 7tr; /Simon Fraser 7cr; /David Parker 12bc; 29bl; /David Parker 31br; /Jean-Loup Charmet 43tc; Hank Morgan 45tr; /Gary Ladd 45crb; 52bl; 53br; /Jean-Loup Charmet 54bl.
Wellcome Institute Library 24crb.
Zefa 41br.

With the exception of the items listed above, and the objects on pages 6, 22, 25, 31c, 33t, 49c, 52, 53t, all the photographs in this book are of objects in the collections of the Science Museum, London.

Glow Stick Art

Alix Wood

PowerKiDS
press

Published in 2020 by Rosen Publishing
29 East 21st Street, New York, NY 10010

Editor: Eloise Macgregor
Designer: Alix Wood

Projects devised and created by Ben Macgregor

Photo Credits: Cover, 1, 8, 9, 14, 15, 16, 17, 20, 21, 22, 23 top, left, and bottom, 24, 25, 26, 27,
28, 29, 32 © Ben Macgregor; 3, 4 bottom, 5, 10, 11, 12, 13, 18, 19 © Alix Wood; 4 top, 6, 7, 23
middle © Adobe Stock Images

Cataloging-in-Publication Data

Names: Wood, Alix.
Title: Glow stick art / Alix Wood.
Description: New York : PowerKids Press, 2020. | Series: Handmade by me |
Includes glossary and index.
Identifiers: ISBN 9781725303027 (pbk.) | ISBN 9781725303041 (library
bound) | ISBN 9781725303034 (6pack)
Subjects: LCSH: Light in art--Juvenile literature. | Handicraft--Juvenile
literature.
Classification: LCC ND1484.W65 2020 | DDC 745.5--dc23

Manufactured in the United States of America

CPSIA Compliance Information: Batch #: CSPK19
For Further Information contact Rosen Publishing, New York, New York at 1-800-237-9932

Contents

Make Cool Glow Stick Gifts

There are so many things you can make using glow sticks. They make great spooky Halloween gifts, and will brighten up any party. You can make lanterns and glow-in-the-dark games, too. Get glowing with these projects!

You Will Need ...

- glow sticks
- **connectors** or drinking straws
- scissors
  - card stock
    - modeling clay
- duct tape and masking tape
- other household items, such as string, boxes, jars, markers, glitter, and ribbon

It is nice to add some spare glow sticks to each gift. Tie a bunch in a bow and wrap them up with your present.

Where Can I Get Glow Sticks?

You can find glow sticks in most craft stores, and even in large grocery stores. If you want to buy a pack of 50 or more, it is usually cheapest to buy them online. Party supply retailers sell them in packs of 50 or 100. The packs usually come with the plastic connectors (right). Perhaps ask an adult to buy some for you online. You will need around 100 to complete all the projects in this book.

Types of Glow Sticks

You can buy several different types of glow sticks. The regular thin ones come in many colors. You can buy longer, wider concert glow sticks, too. High-**intensity** or safety glow sticks are used by the military and glow very brightly. We used these three different types in this book, but you can do all the projects with regular glow sticks.

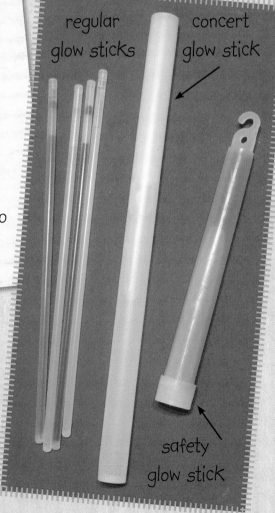

regular glow sticks

concert glow stick

safety glow stick

IMPORTANT - Glow sticks contain chemicals and glass. Be careful not to pierce the sticks when you snap them. The glass and chemicals can irritate the skin.

Getting Started

You can make some really easy gifts using glow sticks. To make a glow-in-the-dark bangle, simply join the two ends of a glow stick together using a plastic connector. You can make a necklace by connecting two glow sticks in the same way.

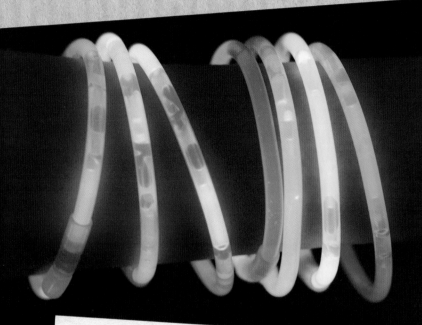

No Connectors? No Problem

You can make your own glow stick connectors using a drinking straw. First, make sure your glow stick fits snugly inside the straw. Then cut the straw into 1-inch (2.5 cm) sections using scissors. You can also use strong tape to join the ends of glow sticks together.

Snapping Glow Sticks

To light a glow stick, gently bend it in the middle. Then shake it. The plastic casing contains a chemical. Inside the casing is a glass tube that holds a second chemical. When you bend a glow stick, you break the glass tube, allowing the two liquids to mix. The **chemical reaction** produces light.

IMPORTANT - Be safe from chemicals and broken glass!
- Never break a glow stick open
- If liquid starts to ooze out, wrap the stick in paper and put it in the trash
- Never put glow sticks into drinks or food
- You cannot recycle glow sticks

MORE IDEAS

Glow sticks can be a really great way to decorate an outdoor nighttime party. Light up the party table by putting a handful of glow sticks into an empty glass. Push a few glow sticks into the soil to light the way down an outdoor path.

Glow Party

Throw a glowing party for a friend or family member as a gift. It is easy to create this amazing party table.

You Will Need ...

- several small glow sticks
- two large glow sticks
- connectors or straws
- a balloon
- an ice bucket
- see-through plates, cups, glasses, and bowls

1

▲ Join several glow sticks into circles using connectors or tape.

2

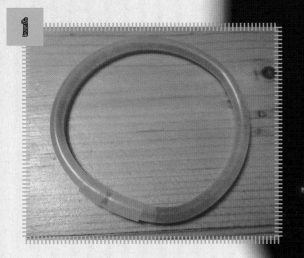

Place single circle glow sticks around, or under, your party glasses. ▶

3

Connect two glow sticks together into a large circle to go under plates. ▲

4

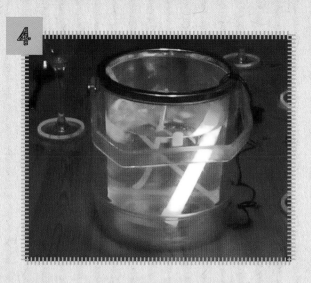

▲ **Put a large glow stick, or several small ones, into an ice bucket, to make a glowing centerpiece.**

Just add cutlery and colorful napkins, and you're ready for the party food!

MORE IDEAS

Decorate your party table with glowing balloons. Inflate a balloon. Quickly push a large glow stick, or a few small ones, through the neck. You may want an adult to help you tie the balloon.

9

Fishing Game

Do you know an **angler**? Try making them this glowing fishing game as a gift. It will certainly make bathtime fun!

You Will Need ...

- glow sticks
- some rubber bands
- a stick
- string
- a paper clip
- some duct tape

1

▲ Find a stick or long twig to be your fishing rod. Trim off any small branches using scissors. You may want an adult to help you.

2

▲ Tie a length of string to the narrowest end of

3

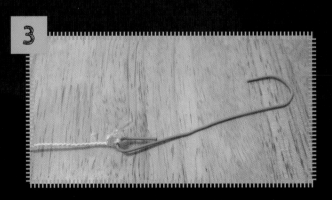

▲ Open out a paper clip to form the hook for your fishing rod. Tie the rod's string to the other end of the paper clip.

▲ To make the handle, wrap some duct tape around the other end of the stick.

▲ When you are ready to play your fishing game, bend the glow sticks into a fish shape, as shown. Secure each fish using a rubber band.

MORE IDEAS

Drop a few more glow sticks in the bathwater, too, to make the water glow.

IMPORTANT - BE SAFE -
If you notice any glow sticks start to leak, get out of the water. The chemicals and glass inside the stick may harm you.

Light-up Saber

Try making this glow stick saber as a gift. If you make two, you can challenge a friend to battle! Simply snap the glow sticks and let battle commence.

You Will Need ...

- a large concert glow stick or several small glow sticks
- cardboard
- rubber bands
- duct tape
- scissors
- tape

1

▲ A large concert glow stick works best for this project.

2

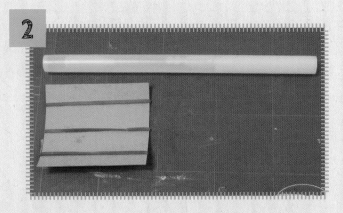

▲ Make a square of cardboard around one-third of the height of your glow stick. Stretch three rubber bands around the square.

MORE IDEAS

If you don't have a large glow stick, you can tape several smaller glow sticks together. Seven sticks of the same color work well. Tape around each end and the middle of the sticks to join them securely together.

3

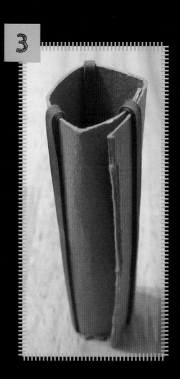

◀ Wrap the cardboard into a cylinder. Tuck the end under a rubber band to hold it in place.

Push the end of your glow stick into the handle. The bands will help to grip the stick. ▼

4

5

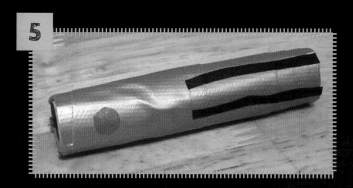

◀ Now cover your handle with silver duct tape. You could cut a red circle and strips of black tape and use them to decorate the handle.

6

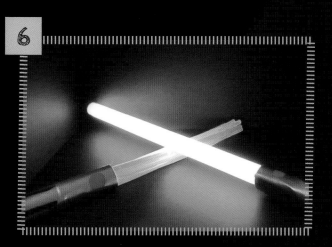

▲ Snap the glow sticks and you are ready to play. When the glow runs out, simply slide a new stick in its place.

Spooky Ghost

This spooky glow stick ghost makes a great Halloween gift. It is really simple to make and looks amazing.

1

▲ Draw some ghostly eyes and a mouth onto a sheet of card stock. Cut them out using scissors.

2

▲ Tape the eyes and mouth near the top of the empty soda bottle.

3

◀ Tie string to the top of a snapped safety glow stick. Place the glow stick in the bottle. Secure it to the neck using modeling clay.

4

▲ Open a large white trash bag. Holding a pencil, push your hand in until you reach the base.

5

▲ Push the pencil through the base of the trash bag to make a hole.

6 Thread the bottle's string through the hole. Tie another piece of string around the top of the bag to keep the glow stick in place.

MORE IDEAS

Make some spooky eyes. Open out a toilet paper tube. Cut out two scary eye shapes. Tape the tube back together.

Place a snapped glow stick in the tube and seal the ends with tape. Place it by your front door to scare your friends!

Bowling Alley

This fun family game should delight someone you know. Glow-in-the-dark bowling adds a new challenge to a popular sport. Try making this simple glow stick project.

You Will Need ...

- six small plastic bottles
- glow sticks
- tape and duct tape
- **acrylic paint** and paintbrush
- plastic ball
- black marker

▲ Cut strips of duct tape to decorate your pins with.

▲ You can paint your lids to match your duct tape strips using acrylic paint.

▲ Cut a plastic ball in half using scissors.

4

▲ Draw some black circles on one half of the ball, to look like finger holes.

5

▲ Once you are ready to play, snap a glow stick and tape it into a small circle. Place it inside half the ball. Tape the ball closed.

6

◀ Snap some more sticks and place one in each bottle, just before you play. Screw on the lids.

MORE IDEAS

You can mark out your alley using more glow sticks.

Zombie Hand Lamp

This spooky light is fun to make. All you need is masking tape, scissors, a straw, and a friend to help you. It is pretty tricky to wrap your own hand, so wrap a friend's hand instead.

1

▲ Place a straw on a friend's wrist as shown. Wrap a length of masking tape, sticky side outward, loosely around their wrist.

You Will Need ...

- bright glow stick
- masking tape
- a straw
- scissors
- a friend to help you

2

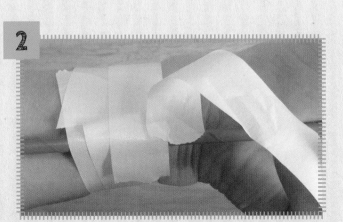

▲ The tape must be sticky side up so it doesn't stick to your friend's skin. It can help to start each piece sticky side down and then turn the tape.

3

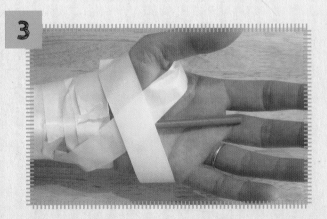

▲ Continue wrapping tape, sticky side out, up the wrist. Then start to wrap around the hand.

18

4

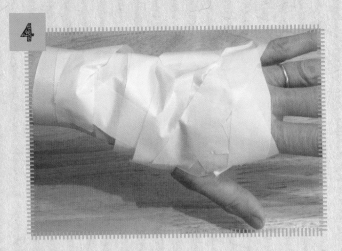

▲ Once the palm is completely covered, put on another layer, this time sticky side down.

5

▲ Now cover the fingers and thumb. Place the tape sticky side up first, then cover each finger sticky side down.

6

◄ Cut along the straw until you can ease your friend's hand out of the covering.

7

▲ Tape along the cut. Pop in a snapped glow stick and your zombie hand lamp is ready!

MORE IDEAS

The glowing zombie hand looks great in a flower bed!

Catherine Wheel

Just fill this spinning wheel with snapped glow sticks, turn out the lights, and spin it!

You Will Need ...

- glow sticks
- scissors
- pencil
- paper
- card stock
- markers
- straws
- paper clip
- **bamboo** pole
- tape

1

▲ Draw around a saucer onto some card stock.

2

▲ Cut out the circle.

3

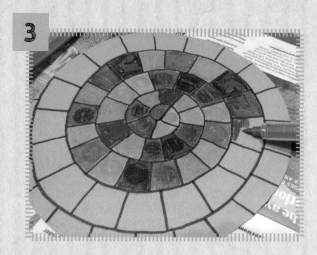

▲ You could draw a spiral design on your card stock circle using markers.

4

▲ Cut some straws into sections. Tape the sections onto the back of your circle, as shown.

5

▲ To find the center of your circle, fold an **identical** paper circle into quarters. Where the folds cross is the center.

6

▲ Pierce the center of the circles using a pencil.

7

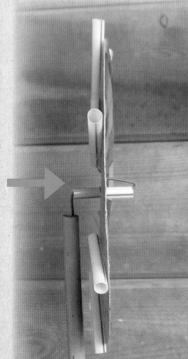

◀ Push an opened paper clip through the center hole. Put a section of straw on either side of the circle. Bend the paper clip at the front and push the back into the bamboo pole. Your wheel should now spin.

Ring Toss Game

Make this ring toss game out of a box, and store all the pieces inside, ready for your next game. Gift it with a handful of glow sticks and connectors.

1

▲ A takeout pizza box is perfect for this project.

You Will Need ...

- a pizza box or similar box
- glow sticks
- connectors or straws
- pencil
- acrylic paint
- paintbrush
- scissors

2

▲ Draw a target on the top of the box, by drawing around different-sized circular objects.

3

▲ Paint the target using a paintbrush and acrylic paint.

4

▲ Pierce a hole in the center of your target using the point of a pencil.

5

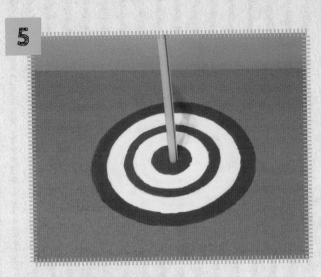

▲ Make the hole large enough and then insert a straw through the hole.

6

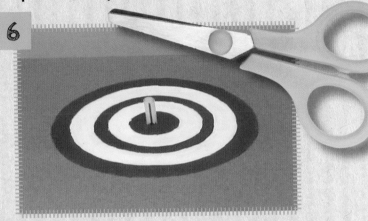

To make the rings, snap and form the glow sticks into a circle. Join the ends using a connector or piece of straw.

▲ Trim the straw so a small section of it sticks up through the top of the pizza box. Put a glow stick in the straw, so it sticks straight up.

MORE IDEAS

If you want, you could just make the rings. Then push a glow stick into the grass and use that as your target!

Chandelier

Make this **chandelier** and brighten up someone's room. To reuse it, simply push new glow sticks into the connectors. We used a safety glow stick as a centerpiece, but ordinary glow sticks work well, too.

You Will Need ...

- cheese triangle box
- glow sticks
- connectors or straws
- shiny gift wrap
- pencil
- scissors
- tape
- string
- duct tape

1

▲ Place the cheese triangle box base onto the gift wrap and draw around it. Cut out the circle.

2

▲ Cut strips of gift wrap the width of the box's sides. Cut enough to go all around the box.

3

▲ Tape the gift wrap circle to the box. Then tape on the sides.

4

▲ Use duct tape to fix the connectors or straw pieces around the inside of the box.

5

▲ Pierce a hole in the center of the box using a pencil. Push some string through the hole. Tie a glow stick to one end.

MORE IDEAS

Make this glow-in-the-dark **coaster**. Glue paper straws to a sheet of card stock. Then slide snapped glow sticks into the straws to make the coaster light up!

6

◄ Push a glow stick into each connector. Then snap the sticks when you are ready to light your chandelier.

Paper Lantern

This beautiful lantern makes an attractive ornament even in daytime. Lit up at night, it looks even more amazing. So many people would love this as a gift!

1

▲ Cut a cross in the center of each side of a square tissue box, to half an inch (1.3 cm) from the edge. Then cut off the triangles to create a square hole.

You Will Need ...

- glow sticks
- a square tissue box
- scissors
- parchment paper
- some leaves and PVA glue
- acrylic paint and paintbrush

2

▲ Paint the box in black acrylic paint.

3

▲ Trace around the box sides onto parchment paper. Cut out four squares and glue a leaf on each.

4

▲ Glue one parchment paper
square inside each cut side,
leaf side facing outward.

5

▲ Push a pen through the
top to help you glue the
far corners of the squares.

6

Drop in a snapped
glow stick when
you want to light
up your lantern.

MORE IDEAS

You can make a simpler
lantern by sticking leaf
shapes cut out of duct tape
onto a plastic milk container.
Drop in a snapped glow stick
and watch it glow.

Snowman Jar

This cute jar makes a lovely gift or decoration for the holiday season. Wrap it up with a handful of glow sticks and it is sure to bring someone joy over many years.

You Will Need ...

- a glow stick
- a jar
- hair spray
- masking tape
- pencil and scissors
- blue and white glitter
- acrylic paint and paintbrush
- a sock
- googly eyes (optional)
- PVA glue (optional)

Lay several strips of masking tape onto a mat. Draw a circle that would fit the side of the jar. Cut out the circle.

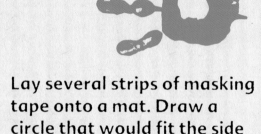

1

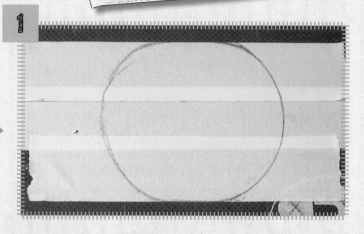

2

▲ Stick the tape circle to the inside of the jar.

3

▲ Go outside and spray hair spray all around the inside of the jar.

4

▲ Sprinkle blue glitter inside the jar. The glitter will stick to the hair spray.

5

▲ Stick on some googly eyes using PVA glue. Or, you can paint eyes using acrylic paint.

6

7

◀ Paint on a carrot nose, rosy cheeks, and a coal mouth using acrylic paint.

▲ Peel off the masking tape. Spray more hair spray into the jar and then sprinkle white glitter where the masking tape used to be.

MORE IDEAS

Make a jar of stars! Punch small holes into a rectangle of aluminum foil. Line the inside of a jar with the foil. Then just add a glow stick.

Glossary

acrylic paint
Fast-drying paint that is water-soluble, but becomes water-resistant when dry.

angler A person who fishes with a hook and line, especially for pleasure.

bamboo A tall, woody grass often with strong, hollow stems.

centerpiece A piece put in the center of something, often as a decoration.

chandelier A branched lighting fixture usually hanging from a ceiling.

chemical reaction A chemical transformation or change.

chemicals Substances obtained from a chemical process.

coaster A shallow container or a plate or mat used to protect a surface.

connectors Devices that join two things.

identical Exactly alike.

intensity Extreme strength or force.

irritate To make sore or inflamed.

Further Information

Baker, Laura. *Cardboard Box Creations*. New York, NY: Lonely Planet Kids, 2018.

Castleforte, Brian. *Papertoy Glowbots: 46 Glowing Robots You Can Make Yourself!* New York, NY: Workman Publishing Company, 2016.

Muldoon, Eilidh. *Gift Boxes to Decorate and Make: For Every Occasion.* London, UK: Nosy Crow, 2018.

Websites

Due to the changing nature of Internet links, PowerKids Press has developed an online list of websites related to the subject of this book. This site is updated regularly. Please use this link to access the list:
www.powerkidslinks.com/hbm/glowstick

Index